LONDON MATHEMATICAL SOCIETY LECTURE NOTE SERIES

Managing Editor:

Professor N. J. Hitchin, Mathematical Institute, University of Oxford, 24-29 St Giles, Oxford OX1 3LB, United Kingdom

The titles below are available from booksellers, or from Cambridge University Press at www.cambridge.org/mathematics

215 Number theory 1992–93, S. DAVID (ed)
216 Stochastic partial differential equations, A. ETHERIDGE (ed)
217 Quadratic forms with applications to algebraic geometry and topology, A. PFISTER
218 Surveys in combinatorics, 1995, P. ROWLINSON (ed)
220 Algebraic set theory, A. JOYAL & I. MOERDIJK
221 Harmonic approximation., S.J. GARDINER
222 Advances in linear logic, J.-Y. GIRARD, Y. LAFONT & L. REGNIER (eds)
223 Analytic semigroups and semilinear initial boundary value problems, KAZUAKI TAIRA
224 Computability, enumerability, unsolvability, S.B. COOPER, T.A. SLAMAN & S.S. WAINER (eds)
225 A mathematical introduction to string theory, S. ALBEVERIO, *et al*
226 Novikov conjectures, index theorems and rigidity I, S. FERRY, A. RANICKI & J. ROSENBERG (eds)
227 Novikov conjectures, index theorems and rigidity II, S. FERRY, A. RANICKI & J. ROSENBERG (eds)
228 Ergodic theory of Z^d actions, M. POLLICOTT & K. SCHMIDT (eds)
229 Ergodicity for infinite dimensional systems, G. DA PRATO & J. ZABCZYK
230 Prolegomena to a middlebrow arithmetic of curves of genus 2, J.W.S. CASSELS & E.V. FLYNN
231 Semigroup theory and its applications, K.H. HOFMANN & M.W. MISLOVE (eds)
232 The descriptive set theory of Polish group actions, H. BECKER & A.S. KECHRIS
233 Finite fields and applications, S. COHEN & H. NIEDERREITER (eds)
234 Introduction to subfactors, V. JONES & V.S. SUNDER
235 Number theory 1993–94, S. DAVID (ed)
236 The James forest, H. FETTER & B. G. DE BUEN
237 Sieve methods, exponential sums, and their applications in number theory, G.R.H. GREAVES *et al*
238 Representation theory and algebraic geometry, A. MARTSINKOVSKY & G. TODOROV (eds)
240 Stable groups, F.O. WAGNER
241 Surveys in combinatorics, 1997, R.A. BAILEY (ed)
242 Geometric Galois actions I, L. SCHNEPS & P. LOCHAK (eds)
243 Geometric Galois actions II, L. SCHNEPS & P. LOCHAK (eds)
244 Model theory of groups and automorphism groups, D. EVANS (ed)
245 Geometry, combinatorial designs and related structures, J.W.P. HIRSCHFELD *et al*
246 *p*-Automorphisms of finite *p*-groups, E.I. KHUKHRO
247 Analytic number theory, Y. MOTOHASHI (ed)
248 Tame topology and o-minimal structures, L. VAN DEN DRIES
249 The atlas of finite groups: ten years on, R. CURTIS & R. WILSON (eds)
250 Characters and blocks of finite groups, G. NAVARRO
251 Gröbner bases and applications, B. BUCHBERGER & F. WINKLER (eds)
252 Geometry and cohomology in group theory, P. KROPHOLLER, G. NIBLO & R. STÖHR (eds)
253 The *q*-Schur algebra, S. DONKIN
254 Galois representations in arithmetic algebraic geometry, A.J. SCHOLL & R.L. TAYLOR (eds)
255 Symmetries and integrability of difference equations, P.A. CLARKSON & F.W. NIJHOFF (eds)
256 Aspects of Galois theory, H. VÖLKLEIN *et al*
257 An introduction to noncommutative differential geometry and its physical applications 2ed,
     J. MADORE
258 Sets and proofs, S.B. COOPER & J. TRUSS (eds)
259 Models and computability, S.B. COOPER & J. TRUSS (eds)
260 Groups St Andrews 1997 in Bath, I, C.M. CAMPBELL *et al*
261 Groups St Andrews 1997 in Bath, II, C.M. CAMPBELL *et al*
262 Analysis and logic, C.W. HENSON, J. IOVINO, A.S. KECHRIS & E. ODELL
263 Singularity theory, B. BRUCE & D. MOND (eds)
264 New trends in algebraic geometry, K. HULEK, F. CATANESE, C. PETERS & M. REID (eds)
265 Elliptic curves in cryptography, I. BLAKE, G. SEROUSSI & N. SMART
267 Surveys in combinatorics, 1999, J.D. LAMB & D.A. PREECE (eds)
268 Spectral asymptotics in the semi-classical limit, M. DIMASSI & J. SJÖSTRAND
269 Ergodic theory and topological dynamics, M.B. BEKKA & M. MAYER
270 Analysis on Lie groups, N.T. VAROPOULOS & S. MUSTAPHA
271 Singular perturbations of differential operators, S. ALBEVERIO & P. KURASOV
272 Character theory for the odd order theorem, T. PETERFALVI
273 Spectral theory and geometry, E.B. DAVIES & Y. SAFAROV (eds)
274 The Mandlebrot set, theme and variations, TAN LEI (ed)
275 Descriptive set theory and dynamical systems, M. FOREMAN *et al*
276 Singularities of plane curves, E. CASAS-ALVERO

277 Computational and geometric aspects of modern algebra, M.D. ATKINSON *et al*
278 Global attractors in abstract parabolic problems, J.W. CHOLEWA & T. DLOTKO
279 Topics in symbolic dynamics and applications, F. BLANCHARD, A. MAASS & A. NOGUEIRA (eds)
280 Characters and automorphism groups of compact Riemann surfaces, T. BREUER
281 Explicit birational geometry of 3-folds, A. CORTI & M. REID (eds)
282 Auslander-Buchweitz approximations of equivariant modules, M. HASHIMOTO
283 Nonlinear elasticity, Y. FU & R.W. OGDEN (eds)
284 Foundations of computational mathematics, R. DEVORE, A. ISERLES & E. SÜLI (eds)
285 Rational points on curves over finite, fields, H. NIEDERREITER & C. XING
286 Clifford algebras and spinors 2ed, P. LOUNESTO
287 Topics on Riemann surfaces and Fuchsian groups, E. BUJALANCE *et al*
288 Surveys in combinatorics, 2001, J. HIRSCHFELD (ed)
289 Aspects of Sobolev-type inequalities, L. SALOFF-COSTE
290 Quantum groups and Lie theory, A. PRESSLEY (ed)
291 Tits buildings and the model theory of groups, K. TENT (ed)
292 A quantum groups primer, S. MAJID
293 Second order partial differential equations in Hilbert spaces, G. DA PRATO & J. ZABCZYK
294 Introduction to the theory of operator spaces, G. PISIER
295 Geometry and Integrability, L. MASON & YAVUZ NUTKU (eds)
296 Lectures on invariant theory, I. DOLGACHEV
297 The homotopy category of simply connected 4-manifolds, H.-J. BAUES
298 Higher operads, higher categories, T. LEINSTER
299 Kleinian Groups and Hyperbolic 3-Manifolds Y. KOMORI, V. MARKOVIC & C. SERIES (eds)
300 Introduction to Möbius Differential Geometry, U. HERTRICH-JEROMIN
301 Stable Modules and the D(2)-Problem, F.E.A. JOHNSON
302 Discrete and Continuous Nonlinear Schrödinger Systems, M. J. ABLORWITZ, B. PRINARI &
     A. D. TRUBATCH
303 Number Theory and Algebraic Geometry, M. REID & A. SKOROBOGATOV (eds)
304 Groups St Andrews 2001 in Oxford Vol. 1, C.M. CAMPBELL, E.F. ROBERTSON &
     G.C. SMITH (eds)
305 Groups St Andrews 2001 in Oxford Vol. 2, C.M. CAMPBELL, E.F. ROBERTSON &
     G.C. SMITH (eds)
306 Peyresq lectures on geometric mechanics and symmetry, J. MONTALDI & T. RATIU (eds)
307 Surveys in Combinatorics 2003, C. D. WENSLEY (ed.)
308 Topology, geometry and quantum field theory, U. L. TILLMANN (ed)
309 Corings and Comodules, T. BRZEZINSKI & R. WISBAUER
310 Topics in Dynamics and Ergodic Theory, S. BEZUGLYI & S. KOLYADA (eds)
311 Groups: topological, combinatorial and arithmetic aspects, T. W. MÜLLER (ed)
312 Foundations of Computational Mathematics, Minneapolis 2002, FELIPE CUCKER *et al* (eds)
313 Transcendental aspects of algebraic cycles, S. MÜLLER-STACH & C. PETERS (eds)
314 Spectral generalizations of line graphs, D. CVETKOVIC, P. ROWLINSON & S. SIMIC
315 Structured ring spectra, A. BAKER & B. RICHTER (eds)
316 Linear Logic in Computer Science, T. EHRHARD *et al* (eds)
317 Advances in elliptic curve cryptography, I. F. BLAKE, G. SEROUSSI & N. SMART
318 Perturbation of the boundary in boundary-value problems of Partial Differential Equations, D. HENRY
319 Double Affine Hecke Algebras, I. CHEREDNIK
321 Surveys in Modern Mathematics, V. PRASOLOV & Y. ILYASHENKO (eds)
322 Recent perspectives in random matrix theory and number theory, F. MEZZADRI &
     N. C. SNAITH (eds)
323 Poisson geometry, deformation quantisation and group representations, S. GUTT *et al* (eds)
324 Singularities and Computer Algebra, C. LOSSEN & G. PFISTER (eds)
325 Lectures on the Ricci Flow, P. TOPPING
326 Modular Representations of Finite Groups of Lie Type, J. E. HUMPHREYS
328 Fundamentals of Hyperbolic Manifolds, R. D. CANARY, A. MARDEN & D. B. A. EPSTEIN (eds)
329 Spaces of Kleinian Groups, Y. MINSKY, M. SAKUMA & C. SERIES (eds)
330 Noncommutative Localization in Algebra and Topology, A. RANICKI (ed)
331 Foundations of Computational Mathematics, Santander 2005, L. PARDO, A. PINKUS, E. SULI &
     M. TODD (eds)
332 Handbook of Tilting Theory, L. ANGELERI HÜGEL, D. HAPPEL & H. KRAUSE (eds)
333 Synthetic Differential Geometry 2ed, A. KOCK
334 The Navier-Stokes Equations, P. G. DRAZIN & N. RILEY
335 Lectures on the Combinatorics of Free Probability, A. NICU & R. SPEICHER
336 Integral Closure of Ideals, Rings, and Modules, I. SWANSON & C. HUNEKE
337 Methods in Banach Space Theory, J. M. F. CASTILLO & W. B. JOHNSON (eds)
338 Surveys in Geometry and Number Theory, N. YOUNG (ed)
339 Groups St Andrews 2005 Vol. 1, C.M. CAMPBELL, M. R. QUICK, E.F. ROBERTSON &
     G.C. SMITH (eds)
340 Groups St Andrews 2005 Vol. 2, C.M. CAMPBELL, M. R. QUICK, E.F. ROBERTSON &
     G.C. SMITH (eds)
341 Ranks of Elliptic Curves and Random Matrix Theory, J. B. CONREY, D. W. FARMER, F.
     MEZZADRI & N. C. SNAITH (eds)

Lectures on Kähler Geometry

ANDREI MOROIANU
École Polytechnique, Paris

CAMBRIDGE
UNIVERSITY PRESS

CAMBRIDGE
UNIVERSITY PRESS

University Printing House, Cambridge CB2 8BS, United Kingdom

Published in the United States of America by Cambridge University Press, New York

Cambridge University Press is part of the University of Cambridge.

It furthers the University's mission by disseminating knowledge in the pursuit of education, learning and research at the highest international levels of excellence.

www.cambridge.org
Information on this title: www.cambridge.org/9780521688970

First published 2007

A catalogue record for this publication is available from the British Library

ISBN 978-0-521-86891-4 Hardback
ISBN 978-0-521-68897-0 Paperback

Contents

Introduction   ix

Part 1.   Basics of differential geometry   1

Chapter 1.   Smooth manifolds   3
  1.1.   Introduction   3
  1.2.   The tangent space   4
  1.3.   Vector fields   6
  1.4.   Exercises   9

Chapter 2.   Tensor fields on smooth manifolds   13
  2.1.   Exterior and tensor algebras   13
  2.2.   Tensor fields   15
  2.3.   Lie derivative of tensors   17
  2.4.   Exercises   19

Chapter 3.   The exterior derivative   21
  3.1.   Exterior forms   21
  3.2.   The exterior derivative   21
  3.3.   The Cartan formula   23
  3.4.   Integration   24
  3.5.   Exercises   26

Chapter 4.   Principal and vector bundles   29
  4.1.   Lie groups   29
  4.2.   Principal bundles   31
  4.3.   Vector bundles   33
  4.4.   Correspondence between principal and vector bundles   33
  4.5.   Exercises   35

Chapter 5.   Connections   37
  5.1.   Covariant derivatives on vector bundles   37
  5.2.   Connections on principal bundles   39
  5.3.   Linear connections   41
  5.4.   Pull-back of bundles   41
  5.5.   Parallel transport   42
  5.6.   Holonomy   43
  5.7.   Reduction of connections   44

5.8.   Exercises   45

Chapter 6.   Riemannian manifolds   47
   6.1.   Riemannian metrics   47
   6.2.   The Levi–Civita connection   48
   6.3.   The curvature tensor   49
   6.4.   Killing vector fields   51
   6.5.   Exercises   52

Part 2.   Complex and Hermitian geometry   55

Chapter 7.   Complex structures and holomorphic maps   57
   7.1.   Preliminaries   57
   7.2.   Holomorphic functions   59
   7.3.   Complex manifolds   59
   7.4.   The complexified tangent bundle   61
   7.5.   Exercises   62

Chapter 8.   Holomorphic forms and vector fields   65
   8.1.   Decomposition of the (complexified) exterior bundle   65
   8.2.   Holomorphic objects on complex manifolds   67
   8.3.   Exercises   68

Chapter 9.   Complex and holomorphic vector bundles   71
   9.1.   Holomorphic vector bundles   71
   9.2.   Holomorphic structures   72
   9.3.   The canonical bundle of $\mathbb{C}P^m$ 74
   9.4.   Exercises   75

Chapter 10.   Hermitian bundles   77
   10.1.   The curvature operator of a connection   77
   10.2.   Hermitian structures and connections   78
   10.3.   Exercises   80

Chapter 11.   Hermitian and Kähler metrics   81
   11.1.   Hermitian metrics   81
   11.2.   Kähler metrics   82
   11.3.   Characterization of Kähler metrics   83
   11.4.   Comparison of the Levi–Civita and Chern connections   85
   11.5.   Exercises   86

Chapter 12.   The curvature tensor of Kähler manifolds   87
   12.1.   The Kählerian curvature tensor   87
   12.2.   The curvature tensor in local coordinates   88
   12.3.   Exercises   91

Chapter 13.   Examples of Kähler metrics   93

13.1.   The flat metric on $\mathbb{C}^m$ 93
13.2.   The Fubini–Study metric on the complex projective space   93
13.3.   Geometrical properties of the Fubini–Study metric   95
13.4.   Exercises   97

Chapter 14.   Natural operators on Riemannian and Kähler manifolds   99
14.1.   The formal adjoint of a linear differential operator   99
14.2.   The Laplace operator on Riemannian manifolds   100
14.3.   The Laplace operator on Kähler manifolds   101
14.4.   Exercises   104

Chapter 15.   Hodge and Dolbeault theories   105
15.1.   Hodge theory   105
15.2.   Dolbeault theory   107
15.3.   Exercises   109

Part 3.   Topics on compact Kähler manifolds   111

Chapter 16.   Chern classes   113
16.1.   Chern–Weil theory   113
16.2.   Properties of the first Chern class   116
16.3.   Exercises   118

Chapter 17.   The Ricci form of Kähler manifolds   119
17.1.   Kähler metrics as geometric U_m-structures   119
17.2.   The Ricci form as curvature form on the canonical bundle   119
17.3.   Ricci-flat Kähler manifolds   121
17.4.   Exercises   122

Chapter 18.   The Calabi–Yau theorem   125
18.1.   An overview   125
18.2.   Exercises   127

Chapter 19.   Kähler–Einstein metrics   129
19.1.   The Aubin–Yau theorem   129
19.2.   Holomorphic vector fields on Kähler–Einstein manifolds   131
19.3.   Exercises   133

Chapter 20.   Weitzenböck techniques   135
20.1.   The Weitzenböck formula   135
20.2.   Vanishing results on Kähler manifolds   137
20.3.   Exercises   139

Chapter 21.   The Hirzebruch–Riemann–Roch formula   141
21.1.   Positive line bundles   141
21.2.   The Hirzebruch–Riemann–Roch formula   142
21.3.   Exercises   145

Chapter 22.   Further vanishing results 147
   22.1.   The Lichnerowicz formula for Kähler manifolds 147
   22.2.   The Kodaira vanishing theorem 149
   22.3.   Exercises 151

Chapter 23.   Ricci-flat Kähler metrics 153
   23.1.   Hyperkähler manifolds 153
   23.2.   Projective manifolds 155
   23.3.   Exercises 156

Chapter 24.   Explicit examples of Calabi–Yau manifolds 159
   24.1.   Divisors 159
   24.2.   Line bundles and divisors 161
   24.3.   Adjunction formulas 162
   24.4.   Exercises 165

Bibliography 167

Index 169

Introduction

These notes grew out of my graduate course at Hamburg University in the autumn of 2003. Their main purpose is to provide a quick and modern introduction to different aspects of Kähler geometry. I had tried to make the original lectures accessible to graduate students in mathematics and theoretical physics having only basic knowledge of calculus in several variables and linear algebra. The present notes should (hopefully) have retained this quality.

The text is organized as follows. The first part is devoted to a review of basic differential geometry. We discuss here topics related to smooth manifolds, tensors, Lie groups, principal bundles, vector bundles, connections, holonomy groups, Riemannian metrics, and Killing vector fields.

The reader familiar with the contents of a first course in differential geometry can pass directly to the second part, which starts with a description of complex manifolds and holomorphic vector bundles. Kähler manifolds are then discussed from the point of view of Riemannian geometry. This part ends with an outline of Hodge and Dolbeault theories, and a simple proof of the famous Kähler identities.

In the third part we study several aspects of compact Kähler manifolds: the Calabi conjecture, Weitzenböck techniques, Calabi–Yau manifolds, and divisors.

The material contained in each chapter is equivalent to a ninety-minute lecture. All chapters end with a series of exercises. Solving them may prove to be at least helpful, if not sufficient, for a reasonable understanding of the theory.

Acknowledgements. I would like to thank Christian Bär and the Department of Mathematics of Hamburg University for having invited me to teach this graduate course. During the preparation of the manuscript I had many discussions with Paul Gauduchon and Uwe Semmelmann which helped me a lot to improve the presentation. I am also indebted to Mihaela Pilca and Liviu Ornea for their critical reading of a preliminary version of these notes. Finally, I would like to thank Roger Astley for his advice and his moral support.

Paris, September 2006

Part 1

Basics of differential geometry

CHAPTER 1

Smooth manifolds

1.1. Introduction

A topological manifold of dimension n is a Hausdorff topological space M which locally "looks like" the space $\mathbb{R}^n$. More precisely, M has an open covering $\mathcal{U}$ such that for every $U \in \mathcal{U}$ there exists a homeomorphism $\phi_U : U \to \tilde{U} \subset \mathbb{R}^n$, called a *local chart*. The collection of all charts is called an *atlas* of M. Since every open ball in $\mathbb{R}^n$ is homeomorphic to $\mathbb{R}^n$ itself, the definition above amounts to saying that every point of M has a neighbourhood homeomorphic to $\mathbb{R}^n$.

EXAMPLES. 1. The sphere $S^n \subset \mathbb{R}^{n+1}$ is a topological manifold of dimension n, with the atlas consisting of the two stereographic projections $\phi_N : S^n \setminus \{S\} \to \mathbb{R}^n$ and $\phi_S : S^n \setminus \{N\} \to \mathbb{R}^n$ where N and S are the North and South poles of S^n.

2. The union $Ox \cup Oy$ of the two coordinate lines in $\mathbb{R}^2$ is not a topological manifold. Indeed, for every neighbourhood U of the origin in $Ox \cup Oy$, the set $U \setminus \{0\}$ has 4 connected components, so U can not be homeomorphic to $\mathbb{R}$.

We now investigate the possibility of defining smooth functions on a given topological manifold M. If $f : M \to \mathbb{R}$ is a continuous function, one is tempted to define f to be smooth if for every $x \in M$ there exists $U \in \mathcal{U}$ containing x such that the composition

$$f_U := f \circ \phi_U^{-1} : \tilde{U} \subset \mathbb{R}^n \to \mathbb{R}$$

is smooth. In order for this to make sense, we need to check that the above property does not depend on the choice of U. Let V be some other element of the open covering $\mathcal{U}$ containing x. If we denote by ϕ_{UV} the *coordinate change* function

$$\phi_{UV} := \phi_U \circ \phi_V^{-1} : \phi_V(U \cap V) \subset \mathbb{R}^n \to \phi_U(U \cap V) \subset \mathbb{R}^n,$$

then

$$f_V = f_U \circ \phi_{UV}.$$

Our definition is thus coherent provided the coordinate changes are all smooth in the usual sense. This motivates the following:

DEFINITION 1.1. *A* smooth (or differentiable) manifold *of dimension n is a topological manifold $(M, \mathcal{U})$ whose atlas $\{\phi_U\}_{U \in \mathcal{U}}$ satisfies the following compatibility condition: for every intersecting $U, V \in \mathcal{U}$, the map between open sets of $\mathbb{R}^n$*

$$\phi_{UV} := \phi_U \circ \phi_V^{-1}$$

is a diffeomorphism. If this condition holds, the atlas $\{\phi_U\}_{U \in \mathcal{U}}$ is also called a smooth structure *on M. An atlas is called* oriented *if the determinant of the Jacobian matrix of ϕ_{UV} is everywhere positive. An* oriented manifold *is a smooth manifold together with an oriented atlas.*

Unless otherwise stated, all smooth manifolds considered in these notes are assumed to be connected.

We have seen that the existence of a smooth structure on M enables one to define smooth functions on M. It is straightforward to extend this definition to functions $f : M \to N$ where M and N are smooth manifolds:

DEFINITION 1.2. *Let $(M, \{\phi_U\}_{U \in \mathcal{U}})$, $(N, \{\psi_V\}_{V \in \mathcal{V}})$ be two smooth manifolds. A continuous map $f : M \to N$ is said to be* smooth *if $\psi_V \circ f \circ \phi_U^{-1}$ is a smooth map in the usual sense for every $U \in \mathcal{U}$ and $V \in \mathcal{V}$. A homeomorphism which is smooth, together with its inverse, is called a* diffeomorphism.

DEFINITION 1.3. *Let M be a smooth manifold. A* local coordinate system *around some $x \in M$ is a diffeomorphism between an open neighbourhood of x and an open set in $\mathbb{R}^n$.*

1.2. The tangent space

From now on, unless otherwise stated, by manifold we understand a smooth manifold with a given smooth structure on it. In order to do calculus on manifolds we need to define objects such as vectors, exterior forms, etc. The main tool for that is provided by the *chain rule*. Recall that if $f : U \subset \mathbb{R}^n \to \mathbb{R}^m$ is a smooth function, its *differential* at any point $x \in U$ is the linear map $df_x : \mathbb{R}^n \to \mathbb{R}^m$ whose matrix in the canonical basis is

$$(df_x)_{ij} := \frac{\partial f_i}{\partial x_j}(x).$$

PROPOSITION 1.4. (Chain rule) *Let $U \subset \mathbb{R}^n$ and $V \subset \mathbb{R}^m$ be two open sets and let $f : U \to \mathbb{R}^m$ and $g : V \to \mathbb{R}^k$ be two smooth maps. Then for every $x \in U \cap f^{-1}(V)$ we have*

$$d(g \circ f)_x = dg_{f(x)} \circ df_x. \tag{1.1}$$

In the particular situation where $m = n = k$, $f : U \to V$ is a diffeomorphism and $g = f^{-1}$, the previous relation reads

$$(d(f^{-1}))_{f(x)} = (df_x)^{-1}. \tag{1.2}$$

Let x be a point of some manifold $(M, \mathcal{U})$ of dimension n. We denote by I_x the set of all $U \in \mathcal{U}$ containing x. On $I_x \times \mathbb{R}^n$ we define the relation "$\sim_x$" by

$$(U, u) \sim_x (V, v) \qquad \Longleftrightarrow \qquad u = (d\phi_{UV})_{\phi_V(x)}(v).$$

By (1.1) and (1.2), "$\sim_x$" is an equivalence relation. An equivalence class is called a *tangent vector* of M at x. By the linearity of the differentials $d\phi_{UV}$ we see that the quotient $I_x \times \mathbb{R}^n / \sim_x$ is an n-dimensional vector space. This vector space is called the *tangent vector space* of M at x and is denoted by $T_x M$. The tangent vector at x defined by the pair (U, u) is denoted by $[U, u]_x$. For each $U \in \mathcal{U}$ containing x, a tangent vector $X \in T_x M$ has a unique representative (U, u) in $\{U\} \times \mathbb{R}^n$. The vector $u \in \mathbb{R}^n$ is the "concrete" representation in the chart ϕ_U of the "abstract" tangent vector X.

DEFINITION 1.5. *The union of all tangent spaces* $TM := \bigsqcup_{x \in M} T_x M$ *is called the* tangent bundle *of* M. *We will see later on that* TM *has a structure of a vector bundle over* M *and is, in particular, a smooth manifold of dimension* $2n$.

If M and N are smooth manifolds, $f : M \to N$ is a smooth map and $x \in M$, one can define the differential $df_x : T_x M \to T_{f(x)} N$ in the following way: choose local charts ϕ_U and ψ_V around x and $f(x)$ respectively and define

$$df_x([U, u]) := [V, d(\psi_V \circ f \circ \phi_U^{-1})_{\phi_U(x)}(u)]. \tag{1.3}$$

Again, the chain rule shows that the definition of df_x does not depend on the choice of the local charts. It is a straightforward exercise in differentials to check the following extension of the chain rule to manifolds:

PROPOSITION 1.6. *Let* M, N *and* Z *be smooth manifolds and let* $f : M \to N$ *and* $g : N \to Z$ *be two smooth maps. Then for every* $x \in M$ *we have*

$$d(g \circ f)_x = dg_{f(x)} \circ df_x.$$

If $f : M \to N$ is a smooth map, the collection $(df_x)_{x \in M}$ defines a map $df : TM \to TN$, called the *differential of* f, which will sometimes be denoted by f_*.

A smooth map $f : M \to N$ is called a *submersion* if its differential df_x is onto for every $x \in M$.

Let M be a smooth manifold of dimension n. A topological subspace $S \in M$ of M is called a *submanifold* of dimension k if for every $x \in S$ there exists a neighbourhood U of x in M and a local coordinate system $\phi_U : U \to \tilde{U}$ such that $S \cap U = \phi_U^{-1}(\mathbb{R}^k \cap \tilde{U})$. The restriction to S of all such coordinate systems provides a smooth structure of dimension k on S.

THEOREM 1.7. (Submersion theorem) *If* $f : M \to N$ *is a submersion then* $f^{-1}(y)$ *is a smooth submanifold of* M *for every* $y \in N$.

PROOF. If y does not belong to the image of f, there is nothing to prove. Otherwise, let $x \in f^{-1}(y)$. By taking local charts around x and y, we can assume that M and N are open subsets of $\mathbb{R}^n$ and $\mathbb{R}^m$ respectively. Since df_x is onto, there exists a non-vanishing $m \times m$ minor in the matrix $(\partial f^i / \partial x_j)_{1 \le i \le m, 1 \le j \le n}$. Without loss of generality we might assume that $(\partial f^i / \partial x_j)_{1 \le i,j \le m}$ is non-zero at x. Consider the map $F : M \to N \times \mathbb{R}^{n-m} \subset \mathbb{R}^n$, $F(z) = (f(z), z_{m+1}, \ldots, z_n)$. The Jacobian of F at x is clearly non-zero, so by the inverse function theorem there exists some open neighbourhood U of x mapped diffeomorphically by F onto some $\tilde{U} \subset \mathbb{R}^n$. By construction $F(f^{-1}(y) \cap U) = (\{y\} \times \mathbb{R}^{n-m}) \cap \tilde{U}$, so we are done.   □

1.3. Vector fields

Let M be a smooth manifold. Every map $X : M \to TM$ such that $X(x) \in T_x M$ for all $x \in M$ defines, for every local chart $\phi : U \to \tilde{U} \subset \mathbb{R}^n$, a map $X_\phi : \tilde{U} \to \mathbb{R}^n$ by

$$X_\phi(\phi(x)) := d\phi_x(X_x).$$

If all these maps are smooth, we say that X is a (smooth) *vector field* on M. For $x \in M$, $X(x)$ (also denoted by X_x) is thus a vector in the tangent space $T_x M$. The set of all vector fields on M is a module over the algebra of smooth functions $\mathcal{C}^\infty(M)$ and is denoted by $\mathcal{X}(M)$.

EXAMPLE. Let e_i denote the constant vector field on $\mathbb{R}^n$ defined by the ith element of the canonical base. If $\phi_U : U \to \tilde{U}$ is a local chart on M, we define the local vector field $\partial / \partial x_i$ on U by $\partial / \partial x_i(x) := [U, e_i]_x$, i.e.

$$(d\phi_U)_x \left(\frac{\partial}{\partial x_i}(x) \right) = e_i \qquad \forall\, x \in U.$$

Since for every $x \in U$, $(d\phi_U)_x$ is (tautologically) an isomorphism between $T_x M$ and $\mathbb{R}^n$, $\{\partial / \partial x_i(x)\}_{i=1,\ldots,n}$ is a basis of $T_x M$. We say that $\{\partial / \partial x_i\}$ is a *local frame* on U.

This notation is motivated by the following:

THEOREM 1.8. *If* $\mathfrak{D}(\mathcal{C}^\infty(M))$ *denotes the Lie algebra of derivations of the algebra of smooth functions on* M, *there exists a natural isomorphism of* $\mathcal{C}^\infty(M)$-*modules* $\Phi : \mathcal{X}(M) \to \mathfrak{D}(\mathcal{C}^\infty(M))$. *In particular,* $\mathcal{X}(M)$ *has a natural Lie algebra structure.*

PROOF. First let $\tilde{U}$ be some open set in $\mathbb{R}^n$. If $\tilde{X} = \sum \tilde{X}^i e_i$ is a smooth vector field on $\tilde{U}$ and $f : \tilde{U} \to \mathbb{R}$ is a smooth function, we define another function $\partial_{\tilde{X}} f$ on $\tilde{U}$ by

$$(\partial_{\tilde{X}} f)(x) := df_x(\tilde{X}_x) = \sum_{i=1}^{n} \frac{\partial f}{\partial x_i}(x) \tilde{X}^i(x).$$

Clearly $\partial_{\tilde{X}} f$ is again smooth, and $\tilde{X}$ defines in this way a derivation of $\mathcal{C}^\infty(U)$:

$$\partial_{\tilde{X}}(fg) = d(fg)(\tilde{X}) = (f\,dg + g\,df)(\tilde{X}) = f\partial_{\tilde{X}}(g) + g\partial_{\tilde{X}} f.$$

If X is now a vector field on a smooth manifold M and $f \in \mathcal{C}^\infty(M)$, we define $(\partial_X f)(x) := df_x(X)$. We will sometimes use the notation $\partial_{X_x} f$ rather than $(\partial_X f)(x)$. Using (1.3) for some chart $\phi_U : U \to \tilde{U}$, one can express the restriction of $\partial_X f$ to U as follows:

$$(\partial_X f)(x) = d(f \circ \phi_U^{-1})_{\phi_U(x)}(\tilde{X}) = (\partial_{\tilde{X}}(f \circ \phi_U^{-1}))(\phi_U(x)), \qquad \forall\, x \in U, \quad (1.4)$$

where $\tilde{X} := d\phi_U(X)$ is the vector field on $\tilde{U}$ corresponding to X in the given chart. The previous argument shows of course that $\partial_X f$ is smooth and that $f \mapsto \partial_X f$ is a derivation.

The map $X \mapsto \partial_X$, clearly defines a morphism $\Phi : \mathcal{X}(M) \to \mathfrak{D}(\mathcal{C}^\infty(M))$ of $\mathcal{C}^\infty(M)$-modules. In order to show that Φ is an isomorphism, we need to make use of so-called *bump functions*. For $x \in \mathbb{R}^n$ and $0 < \varepsilon < \delta$, a smooth function $\varphi : \mathbb{R}^n \to \mathbb{R}$ is called a bump function if it is identically 1 on the open ball $B(x, \varepsilon)$ and vanishes identically outside $B(x, \delta)$. The existence of such functions will be proved in an exercise.

Let $X \in \mathrm{Ker}(\Phi)$. This means that

$$\partial_X f = 0 \qquad\qquad (1.5)$$

for all smooth functions f on M. The (relative) difficulty here is to construct enough smooth functions on M. Let us fix some $x \in M$ and a chart $\phi_U : U \to \tilde{U}$ such that U contains x, and denote by $\tilde{x} := \phi_U(x)$. We choose $0 < \varepsilon < \delta$ such that $\overline{B(\tilde{x}, \delta)} \subset \tilde{U}$ and a bump function φ on $\tilde{U}$ relative to the data $(\tilde{x}, \varepsilon, \delta)$. For every function $\psi : \tilde{U} \to \mathbb{R}$, the function

$$f(y) = \begin{cases} (\varphi\psi)(\phi_U(y)), & y \in U, \\ 0, & y \in M \setminus U, \end{cases}$$

is by construction a smooth function on M. If $\tilde{X} = d\phi_U(X)$ denotes as before the vector field on $\tilde{U}$ which represents X in the chart U, then by (1.4) $\partial_{X_x} f = \partial_{\tilde{X}_{\tilde{x}}}(\varphi\psi)$ and from the properties of φ we obtain

$$\partial_{\tilde{X}_{\tilde{x}}}(\varphi\psi) = \psi(\tilde{x})\partial_{\tilde{X}_{\tilde{x}}}\varphi + \varphi(\tilde{x})\partial_{\tilde{X}_{\tilde{x}}}\psi = \psi(\tilde{x})d\varphi_{\tilde{x}}(\tilde{X}) + d\psi_{\tilde{x}}(\tilde{X}) = d\psi_{\tilde{x}}(\tilde{X}).$$

Using (1.5) we get $d\psi_{\tilde{x}}(\tilde{X}) = 0$ for every smooth function ψ. Taking $\psi = x_i$ shows that $\tilde{X}_i = 0$ at $\tilde{x}$, so finally $X = 0$ at x. Since x was arbitrary, this proves that X vanishes on M.

It remains to show that every derivation D on $\mathcal{C}^\infty(M)$ is defined by a smooth vector field on M. The proof of this fact will be given in an exercise at the end of this chapter. $\qquad\qquad\square$

From now on we will often identify a smooth vector field X and the corresponding derivation $\partial_X : C^\infty(M) \to C^\infty(M)$ on functions. Notice that every tangent vector at some point $X_x \in T_x M$ defines a linear map $\partial_{X_x} : C^\infty(M) \to \mathbb{R}$ which satisfies

$$\partial_{X_x}(fg) = g(x)\partial_{X_x}(f) + f(x)\partial_{X_x}(g), \qquad \forall\, f, g \in C^\infty(M).$$

DEFINITION 1.9. *A path on a manifold M is a smooth map $c : \mathbb{R} \to M$. The tangent vector to c at t, denoted by $\dot{c}(t)$ is by definition the image of the canonical vector $\partial/\partial t \in T_t\mathbb{R}$ through the differential of c at t:*

$$\dot{c}(t) := dc_t\left(\frac{\partial}{\partial t}\right).$$

The formula relating the tangent vector at 0 to a path c, $V := \dot{c}(0)$, and the corresponding linear map ∂_V on functions is

$$\partial_V f = df_{c(0)}(V) = df_{c(0)}\left(dc_0\left(\frac{\partial}{\partial t}\right)\right) = d(f \circ c)_0\left(\frac{\partial}{\partial t}\right) = (f \circ c)'(0). \quad (1.6)$$

DEFINITION 1.10. *Let X be a vector field on M and let x be some point of M. A local integral curve of X through x is a local path $c : (-\varepsilon, \varepsilon) \to M$ such that $c(0) = x$ and $\dot{c}(t) = X_{c(t)}$ for every $t \in (-\varepsilon, \varepsilon)$.*

PROPOSITION 1.11. *Let $X \in \mathcal{X}(M)$ be a smooth vector field on the manifold M.*

(i) For every $x \in M$ there exists $\varepsilon > 0$ such that for every $\delta \le \varepsilon$, there exists a unique integral curve of X through x defined on $(-\delta, \delta)$.

(ii) For every $x \in M$ there exists an open neighbourhood U_x of x and $\varepsilon > 0$ such that the integral curve of X through every $y \in U_x$ is defined for $|t| < \varepsilon$.

(iii) For $x \in M$, let U_x and ε be given by (ii). If $t < \varepsilon$ we define the map $\varphi_t : U_x \to M$ by $\varphi_t(y) := c_y(t)$, where c_y is the integral curve of X through y. Then we have

$$\varphi_t \circ \varphi_s = \varphi_{s+t}, \qquad \forall\, |t|, |s|, |s+t| < \varepsilon, \quad (1.7)$$

on the open set where the composition makes sense.

(iv) For every $t < \varepsilon$, the local map φ_t is a local diffeomorphism.

PROOF. Let $\phi : U \to \tilde{U}$ be a local chart such that $x \in U$ and let $\tilde{X} = d\phi(X)$ be the vector field induced by X on $\tilde{U}$. By Proposition 1.6, a local path c is a local integral curve of X if and only if the local path $\tilde{c} := \phi \circ c$ is a local integral curve of $\tilde{X}$. Since the statement of the proposition is local, we can therefore assume that M is an open subset of $\mathbb{R}^n$. Let us express X and the local path c in the canonical frame of $\mathbb{R}^n$ by $X = \sum X_i e_i$ and

$c(t) = \sum c_i(t)e_i$. The fact that c is an integral curve for X through 0 is equivalent to the following system of ODEs:

$$\begin{cases} c(0) = x, \\ c_i'(t) = X_i(c(t)). \end{cases} \tag{1.8}$$

The statements (i) and (ii) now follow directly from the theorem of Cauchy–Lipschitz.

If $y \in U_x$, both curves $c_1(t) := \varphi_{s+t}(y)$ and $c_2(t) := \varphi_t \circ \varphi_s(y)$ are integral curves of X through $\varphi_s(y)$, so by (i) they coincide. This proves (iii).

The theorem of Cauchy–Lipschitz actually says that the solution of the system (1.8) is a smooth function with respect to both x and t. This shows that each φ_t is smooth. Finally, (iv) follows by taking $t = -s$ in (iii) and using that φ_0 is the identity map by definition. $\qquad\square$

A family of local diffeomorphisms $\{\varphi_t\}$ of M satisfying (1.7) is called a *pseudogroup of local diffeomorphisms* of M. If the local maps φ_t are defined by a vector field X as before, the pseudogroup $\{\varphi_t\}$ defined in Proposition 1.11 is called the *local flow* of X.

A vector field is called *complete* if its flow is globally defined on $M \times \mathbb{R}$.

We will use the flow of vector fields in order to differentiate interesting objects on smooth manifolds called tensor fields. In order to introduce them, we need to recall some background of linear algebra in the next chapter.

1.4. Exercises

(1) Show that a connected topological manifold is path connected.

(2) Use the explicit formulas of the stereographic projections to check that the sphere S^n is a smooth manifold.

(3) Alternatively, show that S^n is a smooth manifold using the submersion theorem.

(4) Show that every connected component of a smooth manifold is again a smooth manifold.

(5) (*The real projective space*) Let $\mathbb{R}P^n$ denote the space of real lines in $\mathbb{R}^{n+1}$ passing through 0. Check that $\mathbb{R}P^n$ is a smooth manifold of dimension n. *Hint:* Denote by U_i the set of lines not contained in the hyperplane $(x_i = 0)$. Show that there exist natural bijections $\phi_i : U_i \to \mathbb{R}^n$ defining a topology and a smooth atlas on $\mathbb{R}P^n$.

(6) If M and N are smooth manifolds, show that $M \times N$ has a smooth structure such that the canonical projections $\pi_M : M \times N \to M$ and $\pi_N : M \times N \to N$ are smooth.

(7) Let M and N be smooth manifolds and let $x \in M$, $y \in N$ be arbitrary points. Show that the tangent space $T_{(x,y)} M \times N$ is naturally isomorphic to $T_x M \oplus T_y N$.

(8) Prove that if $\phi_U : U \to \tilde{U}$ is a local chart on a manifold M and $X = [U, u]$ is an abstract tangent vector at some $x \in U$, then $(d\phi_U)_x(X) = u$. *Hint:* Since $\tilde{U}$ is an open subset of $\mathbb{R}^n$, one may apply (1.3) by choosing the trivial chart $\psi_V = \mathrm{Id}_{\tilde{U}}$.

(9) Let $\{\partial/\partial x_i\}$ and $\{\partial/\partial y_i\}$ be the local frames on M defined by two local charts $\phi : U \to \tilde{U}$ and $\psi : V \to \tilde{V}$. We denote by $\{x_i\}$ and $\{y_i\}$ the coordinates on $\tilde{U}$ and $\tilde{V}$ respectively, and by a slight abuse of notation, we denote by $x = x(y)$ the diffeomorphism $\phi \circ \psi^{-1} : \psi(U \cap V) \to \phi(U \cap V)$. Prove the relation

$$\frac{\partial}{\partial y_i} = \sum_{j=1}^{n} \frac{\partial x_j}{\partial y_i} \frac{\partial}{\partial x_j}$$

between local vector fields on M.

(10) Let φ_t be the local flow of a vector field ξ. Prove that $(\varphi_t)_*(\xi_x) = \xi_{\varphi_t(x)}$ for all t and x where $\varphi_t(x)$ is defined.

(11) Let $f : \mathbb{R} \to \mathbb{R}$ be the smooth function

$$f(t) := \begin{cases} e^{-\frac{1}{t(1-t)}}, & t \in (0,1), \\ 0, & \text{otherwise,} \end{cases}$$

and denote by F the "normalized" primitive of f:

$$F(s) := \frac{\int_s^\infty f(t)dt}{\int_\mathbb{R} f(t)dt}.$$

Prove that for $0 < \varepsilon < \delta$, the function $\varphi : \mathbb{R}^n \to \mathbb{R}$ defined by

$$\varphi(x) := F\left(\frac{|x|^2 - \varepsilon^2}{\delta^2 - \varepsilon^2}\right)$$

is a bump function which is identically 1 on the open ball $B(0,\varepsilon)$ and vanishes identically outside $B(0,\delta)$.

(12) Show that for every derivation D of $C^\infty(M)$ there exists a smooth vector field X such that $D = \partial_X$. *Hint:* Let $p \in M$ and let $(x_1, \ldots, x_n) = \phi_U : U \to \tilde{U}$ be a local coordinate system around p. If φ is a bump function around p with support contained in U, define

$$X_p := \sum_{i=1}^{n} D(\varphi x_i)\left(\frac{\partial}{\partial x_i}\right)_p$$

and show that X_p is independent of φ and ϕ_U. Check the smoothness of the vector field obtained in this way.

(13) Show that every vector field on a compact manifold is complete.

CHAPTER 2

Tensor fields on smooth manifolds

2.1. Exterior and tensor algebras

Let V denote a real vector space of dimension n, with dual vector space denoted by V^*. Let $V^{k,l} := V^{\otimes k} \otimes (V^*)^{\otimes l}$ denote the space of tensors of type (k,l) (we take by convention $V^{\otimes 0} := \mathbb{R}$) and let

$$\mathcal{T}(V) := \bigoplus_{k,l \geq 0} V^{k,l}$$

denote the *tensor algebra* of V. For $k, l \geq 1$ we define the contraction map $C^{k,l} : \mathcal{T}(V) \to \mathcal{T}(V)$ by plugging the vector on the position k into the element of V^* on the position l for each decomposable element. Clearly $C^{k,l}$ vanishes on $V^{k',l'}$ if $k > k'$ or $l > l'$, and maps $V^{k',l'}$ to $V^{k'-1,l'-1}$ otherwise.

Consider the two-sided ideal I of the graded subalgebra

$$\mathcal{T}^0(V) := \bigoplus_{k \geq 0} V^{\otimes k}$$

generated by elements of the form $v_1 \otimes \cdots \otimes v_k$ with $v_i = v_{i+1}$ for some i. It is easy to check that $I = \bigoplus_{k \geq 0} I_k$, where $I_k := I \cap V^{\otimes k}$. Since I is a graded ideal, the quotient $\Lambda V := \mathcal{T}^0(V)/I$ is a graded algebra, called the *exterior algebra*. We denote by $\Lambda^k V := V^{\otimes k}/I_k$ the kth exterior power of V and by π the canonical projection $\pi : V^{\otimes k} \to \Lambda^k V$. The image of $v_1 \otimes \cdots \otimes v_k$ in $\Lambda^k V$ is denoted by $v_1 \wedge \cdots \wedge v_k$. Elements of $\Lambda^k V$ are usually called *multivectors* and elements of $\Lambda^k V^*$ are called *exterior forms*. The tensor product induces a bilinear pairing called *exterior product* on multivectors and on exterior forms.

Recall that there is a canonical isomorphism between tensors of type $(0, p)$ and multilinear forms on V^p which associates to any decomposable element $\alpha_1 \otimes \cdots \otimes \alpha_p \in (V^*)^{\otimes p}$ the multilinear form

$$(\alpha_1 \otimes \cdots \otimes \alpha_p)(v_1, \cdots, v_p) := \prod_{i=1}^{p} \alpha_i(v_i).$$

Similarly, there is a canonical isomorphism between exterior p-forms and multilinear alternate forms on V^p defined on decomposable elements by

$$(\alpha_1 \wedge \cdots \wedge \alpha_p)(v_1, \cdots, v_p) := \det(\alpha_i(v_j)). \tag{2.1}$$

Incidentally, this provides an injective map $\Lambda^p V^* \to (V^*)^{\otimes p}$

$$\alpha_1 \wedge \cdots \wedge \alpha_p \mapsto \sum_{\sigma \in \mathfrak{S}_n} \varepsilon(\sigma) \alpha_{\sigma(1)} \otimes \cdots \otimes \alpha_{\sigma(p)},$$

where $\varepsilon(\sigma)$ denotes the signature of the permutation σ. We will often use this map in order to view exterior p-forms as $(0, p)$-tensors.

If v is a vector and ω is a p-form, viewed as a multilinear alternate map, the *interior product* $v \lrcorner \omega$ is the $(p-1)$-form defined by

$$(v \lrcorner \omega)(v_1, \dots, v_{p-1}) := \omega(v, v_1, \dots, v_{p-1}). \tag{2.2}$$

It is easy to check that this corresponds to

$$v \lrcorner (\alpha_1 \wedge \cdots \wedge \alpha_p) = \sum_{i=1}^{p} (-1)^{i-1} \alpha_i(v) \alpha_1 \wedge \cdots \wedge \alpha_{i-1} \wedge \alpha_{i+1} \wedge \cdots \wedge \alpha_p, \tag{2.3}$$

via the isomorphism (2.1).

LEMMA 2.1. *Let $A : V \to V$ be an automorphism of V. Then A has a unique extension, also denoted by A, to an algebra automorphism of $T(V)$ commuting with all contractions $C^{k,l}$ and preserving the bi-graduation. Moreover this extension preserves the ideal I of the subalgebra $T^0(V^*)$ and thus descends to an automorphism of the exterior algebra ΛV.*

PROOF. We first show the uniqueness of the extension of A to V^*. Since A has to be an algebra automorphism and commutes with $C^{1,1}$ we get for all $v \in V$ and $v^* \in V^*$:

$$\begin{aligned} v^*(v) &= A(v^*(v)) = A(C^{1,1}(v \otimes v^*)) = C^{1,1}(A(v \otimes v^*)) \\ &= C^{1,1}(A(v) \otimes A(v^*)) = A(v^*)(Av), \end{aligned}$$

thus showing that $A(v^*) = v^* \circ A^{-1}$. Since $T(V)$ is generated as algebra by V and V^*, this proves the uniqueness of the extension of A to $T(V)$. In order to prove the existence, we consider the multilinear map $A : V^k \times (V^*)^l \to V^{\otimes k} \otimes (V^*)^{\otimes l}$ defined by

$$A(v_1, \dots, v_k, v_1^*, \dots, v_l^*) := A(v_1) \otimes \cdots \otimes A(v_k) \otimes v_1^* \circ A^{-1} \otimes \cdots \otimes v_l^* \circ A^{-1}.$$

The universality property of the tensor product shows that this map induces a linear endomorphism of $V^{\otimes k} \otimes (V^*)^{\otimes l}$ also denoted by A, which satisfies

$$A(v_1 \otimes \cdots \otimes v_k \otimes v_1^* \otimes \cdots \otimes v_l^*) = A(v_1) \otimes \cdots \otimes A(v_k) \otimes v_1^* \circ A^{-1} \otimes \cdots \otimes v_l^* \circ A^{-1}.$$

It is clear that this extension is an algebra automorphism which commutes with contractions and preserves the ideal I. □

Notice that the above defined extension of automorphisms is natural, in the sense that if A and B are both automorphisms of V, then the extension of $A \circ B$ is the composition of the extensions of A and B.

2.2. Tensor fields

A (smooth) *tensor field* of type (k, l) on an open set of $\mathbb{R}^n$ is a smooth map defined on that open set, with values in $(\mathbb{R}^n)^{k,l}$.

Let now M be a manifold with a smooth atlas $(U, \phi_U)_{U \in \mathcal{U}}$. For $x \in M$ let $T_x^{k,l} M := (T_x M)^{k,l}$ be the space of tensors of type (k, l) defined by the vector space $T_x M$ and let $\mathcal{T}_x M := \bigoplus_{k,l \geq 0} T_x^{k,l} M$ be their direct sum as before. We denote by $\mathcal{T}^{k,l} M$ and $\mathcal{T} M$ the (disjoint) unions $\bigsqcup_{x \in M} T_x^{k,l} M$ and $\bigsqcup_{x \in M} \mathcal{T}_x M$, called the *tensor bundles* of M. Clearly, $\mathcal{T}^{1,0} M$ is just the tangent bundle TM (see Definition 1.5). Its dual, $\mathcal{T}^{0,1} M$, is called the *cotangent bundle* and is usually denoted by $T^* M$.

DEFINITION 2.2. *A* smooth *tensor field* K *of type* (k, l) *on* M *is a map* $K : M \to \mathcal{T}^{k,l} M$ *such that*

- $K(x) \in T_x^{k,l} M$ *for all* $x \in M$ *and*
- K *is smooth in the sense that for every local chart* $\phi : U \to \tilde{U}$, *the map* $d\phi(K) : \tilde{U} \to (\mathbb{R}^n)^{k,l}$ *given by* $d\phi(K)(y) := d\phi_{\phi^{-1}(y)}(K)$ *is smooth.*

The above vector-valued map $d\phi(K)$ *on* $\tilde{U}$ *is called the* local expression *of* K *in the chart* ϕ.

This definition deserves some comments. First, for $x \in U$, the map $d\phi_x$ denotes the extension, given by Lemma 2.1 of the differential $d\phi_x : T_x M \to \mathbb{R}^n$ to a linear map $d\phi_x : T_x^{k,l} M \to (\mathbb{R}^n)^{k,l}$. The fact that the smoothness of $d\phi(K)$ does not depend on the chosen chart ϕ follows from the chain rule, and the naturality property of the extension. Alternatively, one can define a tensor field on M by its coordinates in every local chart, subject to the obvious compatibility conditions on the intersections of charts. Occasionally, tensor fields will be referred to as tensors.

From Lemma 2.1 we see that the algebraic definition of the contraction and of the tensor product on algebraic tensors can be transposed to tensor fields on manifolds. For instance, if K and L are two tensor fields on a manifold M, we define $K \otimes L$ to be the tensor field on M whose local expression in any chart ϕ is the tensor product of the local expressions of K and L in the chart ϕ.

The space of smooth tensors of type (k, l) on M is denoted by $\Gamma(\mathcal{T}^{k,l} M)$ and the direct sum of all these spaces is the algebra of smooth tensor fields $\Gamma(\mathcal{T} M)$. Clearly $\Gamma(\mathcal{T} M)$ is a bi-graded module over $\mathcal{C}^\infty(M)$.

Notice that tensor bundles are special cases of vector bundles, which will be introduced in the next chapter (see Definition 4.6 below). In this framework, a smooth tensor of type (k, l) is just a section of the corresponding vector bundle $\mathcal{T}^{k,l} M$.

A tensor field of type $(1,0)$ is a vector field (i.e. $\mathcal{X}(M) = \Gamma(TM)$), and a tensor field of type $(0,1)$ is called a 1-form. A basic observation in differential geometry is the following:

PROPOSITION 2.3. *The* $C^\infty(M)$-*bilinear pairing*

$$\mathcal{X}(M) \times \Gamma(T^*M) \to C^\infty(M), \qquad (X, \omega) \mapsto \omega(X)$$

defines an isomorphism of $C^\infty(M)$-*modules between* $\Gamma(T^*M)$ *and the dual of* $\mathcal{X}(M)$. *More generally, the* $C^\infty(M)$-*module of smooth tensors of type* (k,l) *is isomorphic to the* $C^\infty(M)$-*module of* $C^\infty(M)$-*multilinear maps defined on* TM^l *with values in* $T^{k,0}M$.

PROOF. It is not easy to understand the full meaning of this statement at first sight. Roughly speaking, it says that if F is a mapping which associates to every smooth vector field a smooth function, and F is $C^\infty(M)$-linear, then the value of $F(X)$ at every point depends on only the value of X at that point. The proof makes use of a localization procedure, similar to Theorem 1.8. One first uses a bump function in order to show that $F(X)_x$ depends on only the value of X in a neighbourhood of x and then computes $F(X)_x$ using a system of local coordinates. Details are left to the reader (cf. [10], p. 26). $\qquad\square$

We now introduce two important notions. If $\varphi : M \to N$ is a diffeomorphism between two manifolds M and N, its differential maps tensor fields of type (k,l) on M to tensor fields of type (k,l) on N by the formula $K \mapsto \varphi_*(K)$, where

$$\varphi_*(K)_y := d\varphi(K_{\varphi^{-1}(y)}).$$

The tensor field $\varphi_*(K)$ is called the *push forward* of K by φ.

If $\varphi : M \to N$ is simply a smooth map, we may define, for every tensor field K of type $(0,l)$ on N, a tensor field f^*K on M, called the *pull-back* of K by φ, by the formula

$$(\varphi^*K)_x(X_1, \ldots, X_l) := K_{\varphi(x)}(d\varphi_x(X_1), \ldots, d\varphi_x(X_l)), \quad \forall\, X_1, \ldots, X_l \in \mathcal{X}(M).$$

Notice that if $\varphi : M \to N$ is a diffeomorphism, the pull-back and the push forward are inverse to each other on covariant tensors:

$$\varphi_* = (\varphi^*)^{-1} = (\varphi^{-1})^*. \tag{2.4}$$

In the particular case where $K = X$ is a vector field on M, $\varphi_*(X)$ is a vector field on N. We would like to compute, for later use, the relationship between the corresponding derivations ∂_X and $\partial_{\varphi_*(X)}$. If $f : N \to \mathbb{R}$ denotes a smooth function and $y \in N$, we have

$$\begin{aligned} \partial_{\varphi_*(X)_y}(f) &= df(\varphi_*(X)_y) = df(d\varphi(X_{\varphi^{-1}(y)})) = d(f \circ \varphi)(X_{\varphi^{-1}(y)}) \\ &= \partial_{X_{\varphi^{-1}(y)}}(f \circ \varphi), \end{aligned}$$

whence
$$\partial_{\varphi_*(X)}(f) = \partial_X(f \circ \varphi) \circ \varphi^{-1}. \tag{2.5}$$

2.3. Lie derivative of tensors

We now introduce the *Lie derivative* of a tensor field K with respect to a vector field X. Let φ_t denote the local flow of X. We define

$$\mathcal{L}_X K := \lim_{t \to 0} \frac{1}{t}\left(K - (\varphi_t)_*(K)\right) = -\frac{d}{dt}\Big|_{t=0} (\varphi_t)_*(K). \tag{2.6}$$

To see that this expression makes sense (remember that it might happen that φ_t are not globally defined for any $t \neq 0$), we notice that for every $x \in M$ there exists some open neighbourhood U of x and some $\varepsilon > 0$ such that φ_t maps U diffeomorphically onto $\varphi_t(U)$ for $t < \varepsilon$, so the limit as $t \to 0$ makes sense at x (we can compute the limit only for t small enough such that $\varphi_{-t}(x) \in U$).

In order to state the main result of this section, we introduce the following notion:

DEFINITION 2.4. *The* Lie bracket *of two vector fields X and Y on a smooth manifold M is the vector field denoted by $[X, Y]$ corresponding (by Theorem 1.8) to the derivation $\partial_X \circ \partial_Y - \partial_Y \circ \partial_X$ on $C^\infty(M)$, that is, the vector field satisfying*

$$\partial_{[X,Y]}(f) = \partial_X(\partial_Y f) - \partial_Y(\partial_X f), \qquad \forall\, f \in C^\infty(M).$$

It is easy to check that $\partial_X \circ \partial_Y - \partial_Y \circ \partial_X$ is indeed a derivation, so the above definition makes sense. Since the Lie bracket clearly satisfies the Jacobi identity $[X, [Y, Z]] + [Y, [Z, X]] + [Z, [X, Y]] = 0$, the space $(\mathcal{X}(M), [.,.])$ is a Lie algebra.

THEOREM 2.5. *Let X be a vector field on M. The Lie derivative $\mathcal{L}_X$ acting on $\Gamma(\mathcal{T}M)$ has the following properties:*

(i) $\mathcal{L}_X$ is a derivation, i.e. $\mathcal{L}_X(K \otimes T) = (\mathcal{L}_X K) \otimes T + K \otimes (\mathcal{L}_X T)$.

(ii) $\mathcal{L}_X$ commutes with the contractions on $\Gamma(\mathcal{T}M)$.

(iii) $\mathcal{L}_X f = \partial_X f$ for all $f \in C^\infty(M)$.

(iv) $\mathcal{L}_X Y = [X, Y]$ for all $Y \in \mathcal{X}(M)$.

(v) (Naturality) If $\varphi : M \to N$ is a diffeomorphism, then

$$\varphi_*(\mathcal{L}_X K) = \mathcal{L}_{\varphi_* X}\varphi_* K, \qquad \forall\, X \in C^\infty(M), \, \forall\, K \in \Gamma(\mathcal{T}M). \tag{2.7}$$

PROOF. Let φ_t be the local flow of X. The first two assertions are straightforward consequences of Lemma 2.1. To prove (*iii*) we use (1.6):

$$\mathcal{L}_X f = -\frac{d}{dt}\Big|_{t=0} (\varphi_t)_*(f) = -\frac{d}{dt}\Big|_{t=0}(f \circ \varphi_{-t}) = \frac{d}{dt}\Big|_{t=0}(f \circ \varphi_t) = \partial_X f.$$

Let now Y be a smooth vector field and let f be a smooth function on M. Using (2.5), (1.6) and the Leibniz rule, we get

$$
\begin{aligned}
\partial_{\mathcal{L}_X Y}(f) &= -\frac{d}{dt}\Big|_{t=0} \partial_{(\varphi_t)_* Y}(f) = -\frac{d}{dt}\Big|_{t=0}\left(\partial_Y(f \circ \varphi_t) \circ \varphi_t^{-1}\right)\\
&= -\left(\frac{d}{dt}\Big|_{t=0}\partial_Y(f \circ \varphi_t)\right) \circ \varphi_0 - \left(\partial_Y(f \circ \varphi_0)\right) \circ \frac{d}{dt}\Big|_{t=0}(\varphi_t^{-1})\\
&= -\partial_Y(\partial_X f) + \partial_X(\partial_Y f) = \partial_{[X,Y]}(f),
\end{aligned}
$$

thus proving (iv). Finally, (v) follows directly from the fact that if φ_t is the local flow of X then $\varphi \circ \varphi_t \circ \varphi^{-1}$ is the local flow of $\varphi_* X$ (see Exercise (7) below):

$$
\begin{aligned}
\mathcal{L}_{\varphi_* X}\varphi_* K &= -\frac{d}{dt}\Big|_{t=0}\varphi_* \circ (\varphi_t)_* \circ \varphi_*^{-1}(\varphi_* K) = -\varphi_*\left(\frac{d}{dt}\Big|_{t=0}(\varphi_t)_* K\right)\\
&= \varphi_*(\mathcal{L}_X K).
\end{aligned}
$$

$\square$

It is sometimes useful to generalize the naturality of the Lie bracket to the case where $\varphi : M \to N$, rather than being a diffeomorphism, is just a smooth map. If X is a vector field on M, $\varphi_* X$ is no longer well-defined, but instead we have the following:

DEFINITION 2.6. *Let $\varphi : M \to N$ be a smooth map. We say that two vector fields $X \in \mathcal{X}(M)$ and $Y \in \mathcal{X}(N)$ are φ-related if $\varphi_*(X_x) = Y_{\varphi(x)}$ for all $x \in M$.*

LEMMA 2.7. (i) *If X and Y are φ-related and $f \in C^\infty(M)$ then*

$$(\partial_Y f) \circ \varphi = \partial_X(f \circ \varphi).$$

(ii) *If X and X' are φ-related to Y and Y' then $[X, X']$ is φ-related to $[Y, Y']$.*

PROOF. (i) For every $x \in M$ we have

$$((\partial_Y f) \circ \varphi)(x) = \partial_{Y_{\varphi(x)}}(f) = \partial_{\varphi_*(X_x)}(f) \stackrel{(2.5)}{=} \partial_{X_x}(f \circ \varphi) = (\partial_X(f \circ \varphi))(x).$$

(ii) Let $x \in M$ and $f \in C^\infty(N)$. Making repeated use of (i) we get

$$
\begin{aligned}
\partial_{[Y,Y']_{\varphi(x)}}(f) &= \partial_{Y_{\varphi(x)}}(\partial_{Y'} f) - \partial_{Y'_{\varphi(x)}}(\partial_Y f)\\
&= \partial_{X_x}((\partial_{Y'} f) \circ \varphi) - \partial_{X'_x}((\partial_Y f) \circ \varphi)\\
&= \partial_{X_x}(\partial_{X'}(f \circ \varphi)) - \partial_{X'_x}(\partial_X(f \circ \varphi))\\
&= \partial_{[X,X']_x}(f \circ \varphi) = \partial_{\varphi_*([X,X']_x)}(f).
\end{aligned}
$$

$\square$

2.4. Exercises

(1) Let $\alpha \in \Lambda^p V$ be a multivector. Show that $\alpha \wedge \alpha = 0$ if p is odd. Does this hold for p even?

(2) If $\alpha \in V$ is a non-zero vector and $\beta \in \Lambda^p V$ is an arbitrary multivector, show that $\alpha \wedge \beta = 0$ if and only if there exists a multivector $\gamma \in \Lambda^{p-1} V$ such that $\beta = \alpha \wedge \gamma$.

(3) If X, Y are vector fields and f, g are functions on M, show that
$$[fX, gY] = fg[X, Y] + f(\partial_X g)Y - g(\partial_Y f)X.$$

(4) If ω is a 1-form, X is a vector field and f is a function, prove that $\mathcal{L}_{fX}\omega = f\mathcal{L}_X\omega - \omega(X)df$.

(5) Prove the following converse to Theorem 2.5: if X is a smooth vector field on a manifold M and $\mathfrak{D}$ is a derivation of $\Gamma(\mathcal{T}M)$ which commutes with contractions and satisfies $\mathfrak{D}f = \partial_X f$ for all $f \in C^\infty(M)$ and $\mathfrak{D}Y = [X, Y]$ for all $Y \in \mathcal{X}(M)$, then $\mathfrak{D} = \mathcal{L}_X$.

(6) Let $\varphi : M \to N$ be a smooth map. Prove the relation
$$\partial_{\varphi_*(X_x)}(f) = \partial_{X_{\varphi(x)}}(f \circ \varphi), \qquad \forall\, X_x \in T_x M. \tag{2.8}$$

(7) Let X be a smooth vector field on M and $\varphi : M \to N$ be a diffeomorphism. Prove that if φ_t denotes the local flow of X, then $\varphi \circ \varphi_t \circ \varphi^{-1}$ is the local flow of $\varphi_* X$ on N.

(8) Let X and Y be arbitrary vector fields. Prove that
$$\mathcal{L}_X \circ \mathcal{L}_Y - \mathcal{L}_Y \circ \mathcal{L}_X = \mathcal{L}_{[X,Y]}. \tag{2.9}$$

Hint: Show that the left hand side term is a derivation of the tensor algebra commuting with contractions, which coincides with the right hand side term on functions and vector fields.

CHAPTER 3

The exterior derivative

3.1. Exterior forms

Let M be a smooth manifold and $x \in M$. A p-form at x is by definition an element of $\Lambda^p(T_x^*M)$, which, we know, is canonically isomorphic to the space of p-linear alternated maps $(T_xM)^p \to \mathbb{R}$. The *exterior bundle* of M is the disjoint union

$$\Lambda^p M := \bigsqcup_{x \in M} \Lambda^p(T_x^*M).$$

A p-form on M is a map $\omega : M \to \Lambda^p M$ such that $\omega_x \in \Lambda^p(T_x^*M)$ for all x. Using the identification between exterior forms and tensors described in Section 2.1, we say that a p-form ω on M is smooth if it is smooth as $(0, p)$-tensor field. From Proposition 2.3 we see that the space of smooth p-forms on M is canonically isomorphic (as $\mathcal{C}^\infty(M)$-module) to the space of alternated $\mathcal{C}^\infty(M)$-multilinear forms on $\mathcal{X}(M)^p$. By (2.1), the explicit isomorphism between the space of smooth exterior forms $\Gamma(\Lambda M)$ and the space of smooth $\mathcal{C}^\infty(M)$-multilinear forms on $\mathcal{X}(M)$ is given on decomposable elements by

$$(\alpha_1 \wedge \cdots \wedge \alpha_p)(X_1, \ldots, X_p) := \det(\alpha_i(X_j)), \qquad \forall\, \alpha_i \in \Gamma(\Lambda^1 M). \tag{3.1}$$

Similarly, one can use (2.2) and (2.3) in order to define the interior product of a vector field X and an exterior form ω:

$$(X \lrcorner \omega)(X_1, \ldots, X_{p-1}) := \omega(X, X_1, \ldots, X_{p-1}),$$

which corresponds to

$$X \lrcorner (\alpha_1 \wedge \cdots \wedge \alpha_p) = \sum_{i=1}^{p} (-1)^{i-1} \alpha_i(X) \alpha_1 \wedge \cdots \wedge \alpha_{i-1} \wedge \alpha_{i+1} \wedge \cdots \wedge \alpha_p,$$

via the isomorphism (3.1). This last relation shows that the interior product is an anti-derivation on ΛM: if θ is a p-form and ω is a q-form, then

$$X \lrcorner (\theta \wedge \omega) = (X \lrcorner \theta) \wedge \omega + (-1)^p \theta \wedge (X \lrcorner \omega). \tag{3.2}$$

3.2. The exterior derivative

From now on we denote the space of smooth p-forms $\Gamma(\Lambda^p M)$ by $\Omega^p M$ and the space of smooth exterior forms $\Gamma(\Lambda M)$ by $\Omega^* M$.

THEOREM 3.1. *Let M be a smooth manifold of dimension n. There exists a unique $\mathbb{R}$-linear endomorphism d of $\Omega^* M$, called the* exterior derivative *or* exterior differential *satisfying the following axioms:*

(i) d maps p-forms to $(p+1)$-forms.

(ii) d is the usual differential on functions as elements of $\Omega^0 M$.

(iii) (Anti-derivation) *If θ is a p-form and ω is a q-form, then*

$$d(\theta \wedge \omega) = d\theta \wedge \omega + (-1)^p \theta \wedge d\omega. \tag{3.3}$$

(iv) $d \circ d = 0$.

(v) (Naturality) *If $\psi : M \to N$ is a smooth map and ω is a smooth form, then $d(\psi^* \omega) = \psi^*(d\omega)$.*

PROOF. Let us first assume that M is an open set of $\mathbb{R}^n$. If x_i denote the standard coordinates on $\mathbb{R}^n$ and for every set $I = \{i_1, \ldots, i_p\}$, $i_1 < \cdots < i_p$, we denote by $dx_I = dx_{i_1} \wedge \cdots \wedge dx_{i_p}$, then every p-form ω can be written $\omega = \sum_I \omega^I dx_I$. There is clearly a unique way of defining $d\omega$ so that the above axioms are satisfied:

$$d\omega = \sum_I d\omega^I \wedge dx_I. \tag{3.4}$$

We have to check that the operator d defined by this formula actually satisfies (i)–(v). (i) and (ii) are clear. Let $\theta = \sum \theta^I dx_I$ and $\omega = \sum \omega^J dx_J$ be forms of degree p and q. We have

$$
\begin{aligned}
d(\theta \wedge \omega) &= \sum_{I,J} d(\theta^I \omega^J) \wedge dx_I \wedge dx_J = \sum_{I,J} (\omega^J d\theta^I + \theta^I d\omega^J) \wedge dx_I \wedge dx_J \\
&= \sum_{I,J} \omega^J d\theta^I \wedge dx_I \wedge dx_J + (-1)^p \sum_{I,J} \theta^I dx_I \wedge d\omega^J \wedge dx_J \\
&= d\theta \wedge \omega + (-1)^p \theta \wedge d\omega,
\end{aligned}
$$

thus proving (iii).

If f is a smooth function,

$$d(df) = d\left(\sum_{i=1}^n \frac{\partial f}{\partial x_i} dx_i \right) = \sum_{i,j=1}^n \frac{\partial^2 f}{\partial x_i \partial x_j} dx_j \wedge dx_i = 0$$

because in the last sum $(\partial^2 f/\partial x_i \partial x_j)$ is symmetric in i and j (by the Schwarz Lemma) and $dx_j \wedge dx_i$ is skew-symmetric. Thus d^2 vanishes on functions, and by (3.4) it vanishes on all exterior forms too, so (iv) holds.

If $\psi : M \to N$ is a smooth map, f is a function on N, X is a vector field on M and $x \in M$, the chain rule yields

$$d(\psi^* f)_x(X) = d(f \circ \psi)_x(X) = df_{\psi(x)} \circ d\psi_x(X) = \psi^*(df)(X),$$

so $d(\psi^* f) = \psi^*(df)$. Using (3.4) again, together with (iii), (iv), and the fact that ψ^* is an algebra morphism with respect to the exterior product, we get

$$
\begin{aligned}
d(\psi^* \omega) &= \sum_I d((\psi^* \omega^I) \psi^* dx_I) = \sum_I d(\psi^* \omega_I) \wedge d\psi^* x_I \\
&= \psi^* \left(\sum_I d\omega^I dx_I \right) = \psi^*(d\omega).
\end{aligned}
$$

This proves (v).

We consider now the general case, where M is a manifold. An exterior form ω on M is by definition a collection of exterior forms ω_U on $\tilde{U}$ for each local chart $\phi_U : U \rightarrow \tilde{U}$, satisfying $\phi^*_{UV} \omega_U = \omega_V$ for each U, V at points where this expression makes sense. The naturality condition then shows that the collection $d\omega_U$ satisfies the same relations, and thus determines a well-defined exterior form denoted $d\omega$ on M. It is straightforward to check that this definition satisfies the required axioms. Finally, the uniqueness of d is a consequence of the naturality axiom. Indeed, it is clear that d is uniquely defined on any coordinate neighbourhood U by $d^U = \phi^*_U \circ d^{\mathbb{R}^n} \circ (\phi^*_U)^{-1}$, and if ω is a form on M and i denotes the inclusion of U into M, then we must have $d\omega|_U = i^*(d\omega) = d(i^*\omega) = d(\omega|_U)$, which means that the restriction of $d\omega$ to U is completely determined. As this holds for every chart U, we have proved the uniqueness of d. □

An exterior form $\omega \in \Omega^p M$ is called *closed* if $d\omega = 0$ and *exact* if there exists some $\tau \in \Omega^{p-1} M$ such that $\omega = d\tau$, Since $d^2 = 0$, every exact form is obviously closed.

3.3. The Cartan formula

LEMMA 3.2. *Let X be some vector field on M. The Lie derivative with respect to X of an exterior form ω is given by the formula*

$$
\mathcal{L}_X \omega = \frac{d}{dt}\Big|_{t=0} \varphi^*_t \omega
$$

and commutes with the exterior derivative.

PROOF. The first assertion follows directly from the definition (2.6), taking (2.4) into account. The second part is a consequence of the naturality of d:

$$
\mathcal{L}_X d\omega = \frac{d}{dt}\Big|_{t=0} \varphi^*_t(d\omega) = \frac{d}{dt}\Big|_{t=0} d(\varphi^*_t \omega) = d(\mathcal{L}_X \omega).
$$

□

We are now ready to prove the following:

THEOREM 3.3. (Cartan formula) *For every vector field X and exterior form ω on M, the following relation holds:*

$$\mathcal{L}_X\omega = d(X \lrcorner \omega) + X \lrcorner d\omega.$$

PROOF. A formal calculation using (3.2) and (3.3) shows that the operator $\omega \mapsto d(X \lrcorner \omega) + X \lrcorner d\omega$ is a derivation of $\Omega^* M$ preserving the degree of forms, just like $\mathcal{L}_X$. It is thus enough to show that they coincide on functions and on 1-forms. For $f \in C^\infty(M)$ we have

$$d(X \lrcorner f) + X \lrcorner df = X \lrcorner df = df(X) = \partial_X f = \mathcal{L}_X f.$$

Moreover, both operators commute with d, so they coincide on exact 1-forms as well. Finally, a localization argument (using bump functions like in the proof of Theorem 1.8) shows that the value of $\mathcal{L}_X\omega$ and $d(X \lrcorner \omega) + X \lrcorner d\omega$ at some point x depends on only the value of ω on a small neighbourhood of x. On such a neighbourhood which is the domain of definition of a local chart, every 1-form can be written $\sum f_i dx_i$. Since $\mathcal{L}_X$ and $d \circ X \lrcorner + X \lrcorner \circ d$ are derivations which coincide on f_i and on dx_i, they have to coincide on all local 1-forms, thus finishing the proof.   □

As an application we have the following explicit formula for the exterior derivative of smooth 1-forms:

COROLLARY 3.4. *If $\omega \in \Omega^1 M$ and $X, Y \in \mathcal{X}(M)$, then*

$$d\omega(X,Y) = \partial_X(\omega(Y)) - \partial_Y(\omega(X)) - \omega([X,Y]).$$

PROOF. From the Cartan formula and Theorem 2.5 we have:

$$
\begin{aligned}
d\omega(X,Y) &= (X \lrcorner d\omega)(Y) = (\mathcal{L}_X\omega)(Y) - d(X \lrcorner \omega)(Y) \\
&= \mathcal{L}_X(\omega(Y)) - \omega(\mathcal{L}_X Y) - \partial_Y(\omega(X)) \\
&= \partial_X(\omega(Y)) - \omega([X,Y]) - \partial_Y(\omega(X)).
\end{aligned}
$$

□

3.4. Integration

The usual theory of integration in $\mathbb{R}^n$ can be easily transposed to smooth manifolds as follows. Let M be a smooth oriented n-dimensional manifold with an oriented atlas $\{\phi_U\}_{U \in \mathcal{U}}$ and let $\omega \in \Omega^n M$ be a compactly supported top degree form on M. Suppose first that the support of ω is contained in some $U \in \mathcal{U}$ and let $f(x)dx_1 \wedge \cdots \wedge dx_n := (\phi_U)_*\omega$ be the expression of ω in the chart (U, ϕ_U). We define

$$\int_M \omega := \int_{\mathbb{R}^n} (\phi_U)_*\omega = \int_{\mathbb{R}^n} f(x)dx_1 \wedge \cdots \wedge dx_n.$$

We need to show that this definition does not depend on the choice of the chart. Suppose that some other $V \in \mathcal{U}$ contains the support of ω, and let

$g(x)dx_1 \wedge \cdots \wedge dx_n := (\phi_V)_*\omega$ be the expression of ω in the chart (V, ϕ_V). We denote by $\phi := \phi_{UV} = \phi_U \circ \phi_V^{-1}$ the coordinate change and by $\psi = \phi^{-1}$ its inverse. Since $\phi_*((\phi_V)_*\omega) = (\phi_U)_*\omega$ we get

$$
\begin{aligned}
f\,dx_1 \wedge \cdots \wedge dx_n &= \phi_*(g\,dx_1 \wedge \cdots \wedge dx_n) \\
&= \psi^*(g(x)dx_1 \wedge \cdots \wedge dx_n) \\
&= (g \circ \psi)d\psi_1 \wedge \cdots \wedge d\psi_n \\
&= (g \circ \psi)\det(J(\psi))dx_1 \wedge \cdots \wedge dx_n,
\end{aligned}
$$

where $J(\psi)$ denotes the determinant of the Jacobian matrix of ψ. Keeping in mind that $\det(J(\psi)) > 0$, the usual change of variable formula in $\mathbb{R}^n$

$$
\int_{\mathbb{R}^n} g\,dx_1 \wedge \cdots \wedge dx_n = \int_{\mathbb{R}^n} (g \circ \psi)|\det(J(\psi))|dx_1 \wedge \cdots \wedge dx_n
$$

shows that $\int_{\mathbb{R}^n}(\phi_U)_*\omega = \int_{\mathbb{R}^n}(\phi_V)_*\omega$. If the support of ω is not contained in a single chart, one has to "chop" it using a *partition of unity* subordinate to $\mathcal{U}$ (see [13], p. 32 for instance): if $\psi_i : M \to \mathbb{R}$ are positive functions, each of them compactly supported in some $U \in \mathcal{U}$, and such that $\sum \psi_i = 1$, we define

$$
\int_M \omega := \sum \int_M \psi_i \omega.
$$

The sum is finite since ω has compact support, and it is straightforward to check that it does not depend on the choice of the partition of unity (ψ_i).

THEOREM 3.5. (The Stokes theorem) *Let M be an oriented n-dimensional manifold and let ω be a smooth $(n-1)$-form with compact support on M. Then*

$$
\int_M d\omega = 0.
$$

PROOF. Consider first the case $M = \mathbb{R}^n$ and $\omega = \sum f_i dx_1 \wedge ..\widehat{dx_i}.. \wedge dx_n$, with support contained in some square $[-N, N]^n$ (where the hat sign means that the corresponding element is missing from the list). Then

$$
\int_{\mathbb{R}^n} d\omega = \sum_{i=1}^{n} \int_{\mathbb{R}^n} \frac{\partial f_i}{\partial x_i}dx_1 \wedge \cdots \wedge dx_n,
$$

and each term in the right hand side sum vanishes because of the Fubini theorem, e.g. for $i = 1$:

$$
\int_{\mathbb{R}^n} \frac{\partial f_1}{\partial x_1} dx_1 \wedge \cdots \wedge dx_n = \int_{[-N,N]^n} \frac{\partial f_1}{\partial x_1} dx_1 \wedge \cdots \wedge dx_n
$$

$$
= \int_{[N,N]^{n-1}} \left(\int_{-N}^{N} \frac{\partial f_1}{\partial x_1} dx_1 \right) dx_2 \wedge \cdots \wedge dx_n
$$

$$
= \int_{[N,N]^{n-1}} (f_1(N, x_j) - f_1(-N, x_j)) dx_2 \wedge \cdots \wedge dx_n
$$

$$
= 0.
$$

The naturality property of the exterior differential shows that the theorem holds true if ω has compact support contained in some coordinate neighbourhood:

$$
\int_M d\omega = \int_{\mathbb{R}^n} (\phi_U)_*(d\omega) = \int_{\mathbb{R}^n} d((\phi_U)_*(\omega)) = 0. \qquad (3.5)
$$

In the general case, we choose a partition of unity (ψ_i) and write

$$
\int_M d\omega = \sum_i \int_M \psi_i d\omega = \sum_i \int_M (d(\psi_i \omega) - d\psi_i \wedge \omega)
$$

$$
\overset{(3.5)}{=} -\sum_i \int_M d\psi_i \wedge \omega = -\int_M \left(d \sum_i \psi_i \right) \wedge \omega = 0,
$$

as $\sum \psi_i = 1$. $\qquad\qquad\qquad\qquad\qquad\qquad\qquad\qquad\qquad \square$

3.5. Exercises

(1) A *distribution* on a smooth manifold is a subbundle of the tangent bundle (see Definition 4.6 below). A vector field X is called tangent to a distribution D if $X_x \in D_x$ for all $x \in M$. The distribution D is called *integrable* if for every pair of vector fields tangent to D, their Lie bracket is also tangent to D. Let α be a nowhere vanishing 1-form on M and $D := \operatorname{Ker}(\alpha)$. Show that D is integrable if and only if $\alpha \wedge d\alpha = 0$.

(2) Let $f : M \to \mathbb{R}$ be a smooth function on a connected manifold M. Show that $df = 0$ if and only if f is constant.

(3) Let $\alpha = \sum \alpha_i dx^i$ be a 1-form on $\mathbb{R}^n$. Show that $d\alpha = 0$ if and only if

$$
\frac{\partial \alpha_i}{\partial x_j} = \frac{\partial \alpha_j}{\partial x_i}, \qquad \forall\, i, j \le n.
$$

(4) (*Poincaré Lemma in degree 1*) Let $\alpha = \sum \alpha_i dx^i$ be a 1-form on $\mathbb{R}^n$. Show that $d\alpha = 0$ if and only if there exists a function $f : \mathbb{R}^n \to \mathbb{R}$ such that $\alpha = df$. *Hint:* If $d\alpha = 0$ define

$$f(x) := \int_0^1 \sum_{i=1}^{n} x_i \alpha_i(tx)\,dt$$

and use the previous exercise and an integration by parts to show that $df = \alpha$.

(5) Use the Cartan formula and an induction on p to show that for every p-form ω and vector fields $X_0, \ldots, X_p$ on a smooth manifold, the following relation holds:

$$d\omega(X_0, \ldots, X_p) = \sum_{i=0}^{p} (-1)^i \partial_{X_i}(\omega(X_0, \ldots, \widehat{X_i}, \ldots, X_p))$$

$$+ \sum_{0 \le i < j \le p} \omega([X_i, X_j], X_0, \ldots, \widehat{X_i}, \ldots, \widehat{X_j}, \ldots, X_p). \tag{3.6}$$

CHAPTER 4

Principal and vector bundles

4.1. Lie groups

A group G which has a smooth manifold structure such that the multiplication $G \times G \to G$ and the inverse map $G \to G$ are smooth is called a *Lie group*.

For $g \in G$ we denote by $L_g : G \to G$ and $R_g : G \to G$ the left and right multiplication by g, called *left and right translations* which are clearly smooth diffeomorphisms of G of inverses $L_{g^{-1}}$ and $R_{g^{-1}}$ respectively. A vector field X on G is called *left invariant* if $dL_g(X) = X$ for every $g \in G$, i.e.

$$(dL_g)_{g^{-1}a}(X_{g^{-1}a}) = X_a, \qquad \forall\, a, g \in G.$$

Taking $g = a$ in this relation shows, in particular, that a left invariant vector field X is completely determined by its value at the identity element e: $X_g = (dL_g)_e(X_e)$. Conversely, this last relation defines a smooth vector field on G for each $X_e \in \mathfrak{g}$. We denote by $\mathfrak{g}$ the vector space of left invariant vector fields on G, called the *Lie algebra* of G. The terminology is justified by Lemma 4.1 below.

Let φ_t denote the local flow of some $X \in \mathfrak{g}$, and let $\varepsilon > 0$ be a positive real number such that the curve $a_t := \varphi_t(e)$ is defined for $t < \varepsilon$. By Proposition 1.11 such an ε exists and a_t is a local integral curve of X, i.e. $\dot{a}_t = X_{a_t}$. If g is any element of G, we have

$$\frac{d}{dt}(ga_t) = \frac{d}{dt}(L_g a_t) = dL_g(\dot{a}_t) = dL_g(X_{a_t}) = X_{ga_t},$$

so ga_t is a local integral curve for X near g. This just means that the local flow of X is $\varphi_t = R_{a_t}$. In particular, (1.7) shows that $a_t a_s = a_{t+s}$ for $t, s, s+t < \varepsilon$, which, of course, allows us to define a_t for every real t by requiring $t \mapsto a_t$ to be a group morphism from $(\mathbb{R}, +)$ to G. The image of this morphism is called the 1-*parameter subgroup* generated by X and one usually denotes $\exp(tX) := a_t$.

LEMMA 4.1. *The space $\mathfrak{g}$ of left invariant vector fields on G is a Lie algebra with respect to the Lie bracket (see Definition 2.4).*

PROOF. We just have to check that $\mathfrak{g}$ is stable under the Lie bracket. This follows from (2.7) applied to left translations: if $X, Y \in \mathfrak{g}$ and $\varphi := L_g$

is a left translation then

$$(L_g)_*[X, Y] = (L_g)_* \mathcal{L}_X Y = \mathcal{L}_{(L_g)_* X} (L_g)_* Y = \mathcal{L}_X Y = [X, Y].$$

□

The above identification $T_e G \simeq \mathfrak{g}$ thus induces a Lie algebra structure on the tangent space at e of G.

LEMMA 4.2. *Let G and H be Lie groups, and let $\varphi : G \to H$ be a Lie group morphism, that is, a smooth map which is also a group morphism. Then the differential at the identity element of φ is a Lie algebra morphism.*

PROOF. If $X \in \mathfrak{g}$ we denote by φX the left invariant vector field on H induced by $\varphi_*(X_e)$. Since φ is a group morphism we have $\varphi \circ L_g = L_{\varphi(g)} \circ \varphi$, whence

$$\varphi_*(X_g) = \varphi_*(L_g)_*(X_e) = (L_{\varphi(g)})_*(\varphi_*(X_e)) = (\varphi X)_{\varphi(g)}, \qquad \forall\, g \in G.$$

This shows that X and φX are φ-related (see Definition 2.6). If $X, Y \in \mathfrak{g}$, Lemma 2.7 shows that $[X, Y]$ is φ-related to $[\varphi X, \varphi Y]$: $\varphi_*([X, Y]_g) = [\varphi X, \varphi Y]_{\varphi(g)}$ for every $g \in G$. Taking $g = e$ yields the result. □

EXAMPLES. 1. The general linear groups with real or complex coefficients $\mathrm{Gl}_n(\mathbb{R})$ and $\mathrm{Gl}_m(\mathbb{C})$. They are both open subsets in the vector spaces $\mathcal{M}_n(\mathbb{R})$ and $\mathcal{M}_m(\mathbb{C})$ of square matrices with real or complex coefficients. The Lie algebra structure on $\mathfrak{gl}_n(\mathbb{R}) = T_{I_n} \mathrm{Gl}_n(\mathbb{R}) \simeq \mathcal{M}_n(\mathbb{R})$ and on $\mathfrak{gl}_m(\mathbb{C}) = T_{I_m} \mathrm{Gl}_m(\mathbb{C}) \simeq \mathcal{M}_m(\mathbb{C})$ is given by $[A, B] = AB - BA$ (exercise).

2. The orthogonal group O_n and the special orthogonal group SO_n

$$\mathrm{O}_n := \{A \in \mathrm{Gl}_n(\mathbb{R}) \mid AA^t = I_n\}, \qquad \mathrm{SO}_n := \{A \in \mathrm{O}_n \mid \det(A) = 1\}.$$

3. The unitary group U_m and the special unitary group SU_m

$$\mathrm{U}_m := \{A \in \mathrm{Gl}_m(\mathbb{C}) \mid A\bar{A}^t = I_m\}, \qquad \mathrm{SU}_m := \{A \in \mathrm{U}_m \mid \det(A) = 1\}.$$

Using the identification of $\mathbb{C}^m$ with $\mathbb{R}^{2m}$ given by

$$(z_1, \ldots, z_m) = (x_1 + iy_1, \ldots, x_m + iy_m) \mapsto (x_1, \ldots, x_m, y_1, \ldots, y_m),$$

we get the inclusions

$$\mathrm{Gl}_m(\mathbb{C}) \subset \mathrm{Gl}_{2m}(\mathbb{R}) \qquad \text{and} \qquad \mathrm{U}_m \subset \mathrm{O}_{2m}.$$

One can also define each of the above groups as the stabilizer of a vector in a certain representation space. For example, O_n is the stabilizer in $\mathrm{Gl}_n(\mathbb{R})$ of the standard Euclidean product on $\mathbb{R}^n$ (viewed as an element of the induced representation of $\mathrm{Gl}_n(\mathbb{R})$ on $(\mathbb{R}^n)^* \otimes (\mathbb{R}^n)^*$).

DEFINITION 4.3. *A group action (to the right) of a Lie group G on a manifold M is a smooth map $M \times G \to M$, denoted by $(m, g) \mapsto mg$ such that*

(i) $me = m, \qquad \forall\, m \in M$;

(ii) $m(gh) = (mg)h, \qquad \forall\, m \in M,\ g, h \in G.$

In particular the maps $R_g : M \to M$ defined by $R_g(m) = mg$ are all diffeomorphisms. The action is called *free* if R_g has no fixed point for all $g \neq e$, and *transitive* if for every $m, m' \in M$ there exists $g \in G$ such that $m' = mg$.

4.2. Principal bundles

Let G be a Lie group and let M be a smooth manifold.

DEFINITION 4.4. *A G-principal bundle (also called a G-structure) over M is a smooth manifold P together with a smooth submersion $\pi : P \to M$ and a group action of G on P to the right which restricts to a free transitive action on each fibre $\pi^{-1}(x)$. G is called the* structure group *of P. A* section *of P is a smooth map $\sigma : M \to P$ such that $\pi \circ \sigma = \mathrm{Id}_M$.*

From the definition we see that each fibre is diffeomorphic to G: just choose some $u \in \pi^{-1}(x)$ and define $L_u : G \to \pi^{-1}(x)$, $g \mapsto ug$. The image of a vector $A \in T_a G$ by the differential of L_u is a vector tangent to the fibre at ua and will be denoted by uA.

It is easy to see that π is an equivariant locally trivial fibration over M with fibre G, in the sense that each point $x \in M$ has a neighbourhood U and a diffeomorphism $\psi : \pi^{-1}U \to U \times G$ which is G-equivariant, that is, $\psi(y, gh) = \psi(y, g)h$. To prove this, we apply the submersion theorem 1.7 in order to construct a local section σ of P in a neighbourhood of x and define $\psi^{-1}(y, g) := \sigma(y)g$. We can construct in this way an open cover $\{(U_\alpha)_{\alpha \in I}\}$ of M with trivializations $\psi_\alpha : \pi^{-1}U_\alpha \to U_\alpha \times G$ and a Čech cocycle $\{(\psi_{\alpha\beta})_{\alpha,\beta \in I}\}$ relative to this open cover where $\psi_{\alpha\beta} : U_\alpha \cap U_\beta \to G$ is defined by $\psi_\alpha \circ \psi_\beta^{-1}(x, g) = (x, \psi_{\alpha\beta}(x)g)$. For the definition and properties of Čech cocycles and cohomology see [3], p. 34.

Conversely, given a Čech cocycle $\{(U_\alpha)_{\alpha \in I},\ \psi_{\alpha\beta} : U_\alpha \cap U_\beta \to G\}$ (so that $\psi_{\alpha\beta} \circ \psi_{\beta\gamma} \circ \psi_{\gamma\alpha} = e$ on $U_\alpha \cap U_\beta \cap U_\gamma$), consider the equivalence relation $\sim$ on the disjoint union $\tilde{P} := \bigsqcup_\alpha (U_\alpha \times G)$,

$$(\alpha, x, g) \sim (\beta, y, h) \iff x = y \text{ and } g = \psi_{\alpha\beta}(x)h.$$

It is easy to check that the quotient $P := \tilde{P}/\!\sim$ is a G-principal bundle over M and the two constructions above are inverse to each other.

If $P(G, M)$ and $Q(G', M')$ are two principal bundles, a *bundle morphism* is a pair (F, f) consisting in a smooth map $F : P \to Q$ and a Lie group morphism $f : G \to G'$ such that $F(ug) = F(u)f(g)$ for all $u \in P$ and $g \in G$. In particular F preserves the fibres and thus defines a smooth map $M \to M'$. P and Q are said to be *isomorphic* if there exist two morphisms between them inverse to each other. Two G-principal bundles are isomorphic if and only if any Čech cocycles defining them induce the same element in Čech cohomology.

EXAMPLES. 1. The trivial principal bundle $P = M \times G$.

2. The frame bundle. If M is a manifold of dimension n, a *frame* at $x \in M$ is a linear isomorphism $u : \mathbb{R}^n \to T_x M$. Let $\mathrm{Gl}_x(M)$ denote the set of all frames at x and let $\mathrm{Gl}(M)$ be the union of all $\mathrm{Gl}_x(M)$. Define $\pi : \mathrm{Gl}(M) \to M$ by associating to a frame u its foot point x. Every local chart on M, $\phi_U : U \to \tilde{U} \subset \mathbb{R}^n$, induces a bijection $\psi_U : \pi^{-1}U \to \tilde{U} \times \mathrm{Gl}_n(\mathbb{R})$ by $\psi_U(u_x) = (\phi_U(x), (d\phi_U)_x \circ u_x)$. Clearly the transition functions

$$\psi_U \circ \psi_V^{-1} = (\phi_U \circ \phi_V^{-1}, d(\phi_U \circ \phi_V^{-1}))$$

are diffeomorphisms on their open sets of definition, so $\mathrm{Gl}(M)$ has a smooth structure which turns π into a smooth submersion. The general linear group $\mathrm{Gl}_n(\mathbb{R})$ (viewed as the automorphism group of $\mathbb{R}^n$) acts on $\mathrm{Gl}(M)$ by $(u, g) \mapsto u \circ g$ and this action restricts to a free transitive action on each fibre $\mathrm{Gl}_x(M)$.

DEFINITION 4.5. *Let $\pi : P \to M$ be a G-structure and let $f : H \to G$ be a (not necessarily injective) Lie group morphism. A reduction (relative to f) of the structure group of P to H is an H-principal bundle Q and a bundle morphism $Q \to P$ whose associated Lie group morphism is f.*

EXAMPLES. 1. If $H \subset G$ is a subgroup of G, a reduction of the structure group of P to H relative to the inclusion of H in G is a subset Q of P such that $\pi : Q \to M$ is an H-principal bundle over M.

2. An *orientation* on a smooth manifold is a reduction of the frame bundle to $\mathrm{Gl}_n^+(\mathbb{R})$.

3. A spin structure on an oriented Riemannian manifold is a reduction of the structure group of the SO_n-structure to Spin_n relative to the canonical covering morphism $\mathrm{Spin}_n \to \mathrm{SO}_n$.

There exists a reduction of the structure group of P to H relative to f if and only if one can find a Čech cocycle $g_{\alpha\beta}$ on M with values in H such that $f(g_{\alpha\beta})$ is a Čech cocycle representing P.

Conversely, one can always *enlarge* the structure group of a G-structure P relative to a Lie group morphism $G \to K$, by composing with f any Čech cocycle representing P. These two operations are inverse to each other (up to isomorphisms).

4.3. Vector bundles

DEFINITION 4.6. *Let M be a smooth manifold. A rank k vector bundle over M is a smooth manifold E together with a submersion $\pi : E \to M$ such that*

(i) Each fibre $E_x := \pi^{-1}(x)$ has a structure of k-dimensional real vector space.

(ii) (Local triviality) For every $x \in M$ there exists an open neighbourhood U of x and a diffeomorphism $\psi_U : \pi^{-1}U \to U \times \mathbb{R}^k$ whose restriction to E_y is a vector space isomorphism onto $\{y\} \times \mathbb{R}^k$ for every $y \in U$.

A subbundle of E is a vector bundle F over M such that F_x is a vector subspace of E_x for every $x \in M$.

A section of E is a smooth map $\sigma : M \to E$ such that $\pi \circ \sigma = \mathrm{Id}_M$. The space of all sections of E is denoted by $\Gamma(E)$.

The most important examples of vector bundles over M are the tensor bundles $\mathcal{T}^{k,l}M$. Indeed, every local chart $\phi_U : U \to \mathbb{R}^n$ of M, induces a local trivialization $\psi_U : \mathcal{T}^{k,l}M|_U \to U \times (\mathbb{R}^n)^{k,l}$ defined by

$$\psi_U(K_x) = (x, (\psi_U)_*(K_x)), \qquad \forall\, x \in U,\ \forall\, K_x \in \mathcal{T}^{k,l}_x M.$$

As before, a rank k vector bundle can also be defined by a Čech cocycle $\{(U_\alpha)_{\alpha \in I},\ \psi_{\alpha\beta} : U_\alpha \cap U_\beta \to \mathrm{Gl}_k(\mathbb{R})\}$ as the quotient space $\bigsqcup_\alpha (U_\alpha \times \mathbb{R}^k)/_\sim$ where $\sim$ is the equivalence relation

$$(\alpha, x, \xi) \sim (\beta, y, \zeta) \iff x = y \text{ and } \xi = \psi_{\alpha\beta}(x)\zeta$$

on the disjoint union $\bigsqcup_\alpha (U_\alpha \times \mathbb{R}^k)$.

Notice that a Čech cocycle with values in a general linear group defines simultaneously a principal bundle (as shown in the previous section) and a vector bundle. The two bundles obtained in this way are of course strongly related, a fact which we now explain.

4.4. Correspondence between principal and vector bundles

To any G-principal bundle P over a manifold M and k-dimensional representation $\rho : G \to \mathrm{Gl}_k(\mathbb{R})$ of G one can associate a rank k vector bundle $E := P \times_\rho \mathbb{R}^k := P \times \mathbb{R}^k/_\sim$ where $\sim$ is the equivalence relation on $P \times \mathbb{R}^k$

$$(u, \xi) \sim (v, \zeta) \iff \exists g \in G \text{ such that } v = ug \text{ and } \zeta = \rho(g^{-1})\xi. \qquad (4.1)$$

The equivalence class defined by a pair $(u, \xi) \in P \times \mathbb{R}^k$ is denoted by $[u, \xi]$, and has to be understood as the element in E represented by ξ in the frame u. By (4.1) we have $[ug, \xi] = [u, \rho(g)\xi]$, which means that ξ represents the same element of E in the frame ug as $\rho(g)\xi$ in the frame u. The fact that E is indeed a vector bundle can be easily checked using a local trivialization of P. Alternatively, one can notice that if P is defined by a Čech cocycle $\{(U_\alpha)_{\alpha \in I}, \ \psi_{\alpha\beta} : U_\alpha \cap U_\beta \to G\}$, then E is the vector bundle defined by the cocycle $\{(U_\alpha)_{\alpha \in I}, \ \rho \circ \psi_{\alpha\beta}\}$. The vector bundle E is called the *associated vector bundle* to (P, ρ).

REMARK. More generally, if ρ denotes a Lie group action of G on a smooth manifold F, we can define a locally trivial fibration $P \times_\rho F$ over M with fibre F using an equivalence relation similar to (4.1).

Conversely, if E is a rank k vector bundle, we define the frame bundle $\mathrm{Gl}(E)$, whose fibre at x is the set of all isomorphisms $u : \mathbb{R}^k \to E_x$. $\mathrm{Gl}(E)$ is clearly a $\mathrm{Gl}_k(\mathbb{R})$-principal bundle over M with respect to the group action $\mathrm{Gl}(E) \times \mathrm{Gl}_k(\mathbb{R}) \to \mathrm{Gl}(E)$ defined by $(u, A) \mapsto u \circ A$. Again, one can easily check that $\mathrm{Gl}(E)$ is represented as $\mathrm{Gl}_k(\mathbb{R})$-principal bundle over M by the very same Čech cocycles which represent E as rank k vector bundle.

In particular (by taking $G = \mathrm{Gl}_k(\mathbb{R})$ and $\rho = \mathrm{Id}$ in the above construction), we see that the two operations $P \mapsto E := P \times_{\mathrm{Id}} \mathbb{R}^k$ and $E \mapsto P := \mathrm{Gl}(E)$ between $\mathrm{Gl}_k(\mathbb{R})$-principal bundles and rank k vector bundles are inverse to each other.

Let P be a G-principal bundle over M, let ρ be a representation of G on a vector space V and let $E = P \times_\rho V$ be the associated vector bundle to (P, ρ). We denote by H the stabilizer of some element of V

$$H := \{g \in G \mid \rho(g)\xi = \xi\}$$

and by $\mathcal{O}$ the G-orbit of ξ in V. The associated fibration $P \times_\rho \mathcal{O}$ is obviously a subset of E, which we will denote by E^ξ.

PROPOSITION 4.7. *There is a canonical one-to-one correspondence between reductions to H of the G-structure P and sections of E^ξ.*

PROOF. If Q denotes a reduction to H of P, we define $\sigma(x) := [u_x, \xi]$, where u_x denotes any element of Q_x. This does not depend on the choice of u_x: if v_x is some other element in the fibre Q_x and $h \in H$ denotes the element of H such that $v_x = u_x h$ then

$$[v_x, \xi] = [u_x h, \xi] = [u_x, \rho(h)\xi] = [u_x, \xi].$$

Conversely, if σ is a section of E^ξ, we define

$$Q_x = \{u_x \in P_x \mid [u_x, \xi] = \sigma(x)\}. \tag{4.2}$$

It is easy to check that H acts freely and transitively on Q_x provided Q_x is non-empty! This last fact is guaranteed by the fact that σ is a section of E^ξ: $\sigma(x) = [u_x, \zeta]$ for some $\zeta \in \mathcal{O}$, so by definition there exists $g \in G$ such that $\zeta = \rho(g)\xi$ and thus $\sigma(x) = [u_x, \rho(g)\xi] = [u_x g, \xi]$. $\square$

EXAMPLE. Let M be a smooth manifold of dimension n. A Riemannian metric (i.e. a smooth section of the bundle of symmetric bilinear forms which is positive definite at each point) corresponds to a reduction to O_n of the frame bundle $\mathrm{Gl}(M)$.

4.5. Exercises

(1) Show that the identification of T_eG with the space of right invariant vector fields induces the opposite Lie algebra structure. *Hint:* Consider the Lie group $\bar{G}$ equal to G as smooth manifold, with the multiplication $g\bar{\cdot}h := hg$ and apply Lemma 4.2 to the Lie group morphism $G \to \bar{G}$, $g \mapsto g^{-1}$.

(2) Prove that the Lie algebra structure on $\mathfrak{gl}_n(\mathbb{R}) = T_{I_n}\mathrm{Gl}_n(\mathbb{R}) \simeq M_n(\mathbb{R})$ is given by $[A, B] = AB - BA$.

(3) Prove that O_n is a Lie group. *Hint:* Apply the submersion theorem to the map $f : \mathrm{Gl}_n(\mathbb{R}) \to S_n$, $f(A) = AA^t$, where S_n denotes the vector space of symmetric matrices.

(4) Prove that if a G-principal bundle over M admits a reduction to the trivial subgroup $\{e\}$ of G, then it is isomorphic to the trivial bundle $M \times G$. Equivalently, a principal bundle is isomorphic to the trivial bundle if and only if it admits a global section.

(5) Show that a G-principal bundle admits a reduction to some subgroup H of G if and only if there exists a Čech cocycle $\{(U_\alpha)_{\alpha \in I}, \ \psi_{\alpha\beta}\}$ defining it such that $\psi_{\alpha\beta}$ take values in H.

CHAPTER 5

Connections

5.1. Covariant derivatives on vector bundles

A smooth function with values in $\mathbb{R}^k$ on a manifold M can be viewed as a section of the trivial vector bundle $M \times \mathbb{R}^k$. The theory of connections is an attempt to generalize the notion of directional derivative of (real or vector-valued) functions to sections in vector bundles.

Let $\pi : E \to M$ be a vector bundle. We are interested in operators which assign to each smooth vector field X on M and smooth section σ of E another smooth section of E called the *covariant derivative* of σ with respect to X. Of course, we would like these operators to be $\mathbb{R}$-linear, tensorial in the first variable and to satisfy the Leibniz rule. Summarizing, we have:

DEFINITION 5.1. *A covariant derivative on E is an $\mathbb{R}$-linear operator* $\nabla : C^\infty(M) \times \Gamma(E) \to \Gamma(E)$ *denoted by* $(X, \sigma) \mapsto \nabla_X \sigma$ *such that for all* $f \in C^\infty(M)$, $X \in \mathcal{X}(M), \sigma \in \Gamma(E)$ *the following conditions are satisfied:*

(i) *(Tensoriality)* $\nabla_{fX}\sigma = f\nabla_X\sigma$.

(ii) *(Leibniz rule)* $\nabla_X(f\sigma) = f\nabla_X\sigma + (\partial_X f)\sigma$.

The first condition simply says that given a section σ, the value of $\nabla_X \sigma$ at some $p \in M$ depends on only the value of X at p (Proposition 2.3).

A covariant derivative is of course not tensorial in its second argument, but the Leibniz rule shows that it is nonetheless *local*, in the sense that for every $p \in M$, if two sections σ and σ' of E coincide on some neighbourhood U of p, then $(\nabla_X \sigma)(p) = (\nabla_X \sigma')(p)$. To see this, take a bump function f which equals 1 in some neighbourhood of p and vanishes outside U and apply the Leibniz rule:

$$(\nabla_X(f\sigma))(p) = f(p)(\nabla_X\sigma)(p) + (\partial_{X_p}f)\sigma(p) = (\nabla_X\sigma)(p).$$

Since by construction $f\sigma = f\sigma'$ everywhere on M, we are done.

We actually have a much stronger statement:

LEMMA 5.2. *Let $p \in M$ and $X \in T_pM$. Assume that E is a vector bundle over M with covariant derivative ∇. If two sections σ and σ' of E satisfy $\sigma(p) = \sigma'(p)$, then $d\sigma_p(X) = d\sigma'_p(X)$ if and only if $\nabla_X\sigma = \nabla_X\sigma'$ at p. In* *other words, the covariant derivative only depends on the 1-jet of the section at any point.*

PROOF. It is enough to show that if a section σ satisfies $\sigma(p) = 0$ then $d\sigma_p(X) = 0$ if and only if $(\nabla_X\sigma)(p) = 0$. Since ∇ is a local operator, we can work in a local trivialization of E over some neighbourhood U of p, providing a diffeomorphism $\psi : \pi^{-1}U \to M \times \mathbb{R}^k$. Let us write $\psi \circ \sigma = (f_1, \dots, f_k)$, where f_i are real functions on U. The hypothesis on σ reads $f_i(p) = 0$. If we denote by σ_i the local basis of E over U corresponding to the canonical basis of $\mathbb{R}^k$ via ψ, we have $\sigma = \sum f_i\sigma_i$, so by the Leibniz rule

$$\nabla_{X_p}\sigma = \sum f_i(p)\sigma_i(p) + \sum(\partial_{X_p}f_i)\sigma_i(p) = \sum(\partial_{X_p}f_i)\sigma_i(p).$$

We thus have

$$\nabla_{X_p}\sigma = 0 \iff \partial_{X_p}f_i = 0 \; \forall \, i \iff (\psi \circ \sigma)_*(X_p) = 0 \iff \sigma_*(X_p) = 0.$$

$\square$

Given a vector $X \in T_pM$, a local section σ is called *parallel* in the direction of X at p if $\nabla_{X_p}\sigma = 0$. More generally, σ is called parallel along a curve γ_t in M if $\nabla_{\dot\gamma_t}\sigma = 0$ for all t.

LEMMA 5.3. *For every $X \in T_pM$ and $\sigma_0 \in E_p$ there exists a local section σ of E which is parallel in the direction of X at p and satisfies $\sigma(p) = \sigma_0$.*

PROOF. Let (x_i) be a local system of coordinates on some neighbourhood U of p, and let σ_j be a local basis of E around p given by a local trivialization of E as before. We take some arbitrary local section τ of E around p such that $\tau(p) = \sigma_0$. There exist real numbers $a_1, \dots, a_n$ and $b_1, \dots, b_k$ such that $X = \sum a_i(\partial/\partial x_i)(p)$ and $\nabla_X\tau = \sum b_j\sigma_j(p)$. If $X = 0$ there is nothing to prove, so assume for example that $a_1 \neq 0$. It is then straightforward to check that $\sigma := \tau - (x_1/a_1)\sum b_j\sigma_j$ satisfies $\sigma(p) = \sigma_0$ and $\nabla_X\sigma = 0$. $\square$

LEMMA 5.4. *If σ is a parallel section along some curve γ_t which vanishes at γ_0, then σ vanishes identically on γ.*

PROOF. Let I be the interval of definition of γ. By Lemma 5.2 we must have $d\sigma(\dot\gamma_t) = 0$ for all $t \in I$, which means that the differential of the smooth map $I \to E$, given by $t \mapsto \sigma(\gamma_t)$ vanishes. Therefore this map is constant, equal to $\sigma(\gamma_0) = 0$. $\square$

Using Lemma 5.2 we obtain for every $e \in E$ a linear map $h : T_{\pi(e)}M \to T_eE$, denoted by $X \mapsto \tilde{X} := \sigma_*(X)$, where σ is any local section of E which is parallel in the direction of X at p and satisfies $\sigma(p) = e$. By construction we have

$$\pi_* \circ h = \text{Id}. \tag{5.1}$$

The vector $\tilde{X}$ is called the *horizontal lift* of X at e and the image of h, denoted by $T_e^h E$ is called the *horizontal subspace* of $T_e E$. Since π_* is onto, (5.1) shows that h is an isomorphism from $T_{\pi(e)} M$ to $T_e^h E$. If we denote the tangent space to the fibre of E at e by $T_e^v E$, (5.1) also yields that $T_e E = T_e^v E \oplus T_e^h E$. The covariant derivative ∇ defines in this way a splitting of the exact sequence of vector bundles over E

$$0 \to T^v E \to TE \to \pi^* TM \to 0,$$

where $\pi^* TM$ is the pull-back by $\pi : E \to M$ of the tangent bundle of M (see Section 5.4 below). Using local trivializations of E and the fact that ∇ is smooth, it is straightforward to show that the union of all subspaces $T_e^h E$ forms a smooth vector subbundle of rank n of TE, called the *horizontal distribution*.

5.2. Connections on principal bundles

We will now interpret the covariant derivative ∇ on E in terms of the frame bundle $\mathrm{Gl}(E)$. Recall that for $x \in M$, an element $u \in \mathrm{Gl}_x(E)$ is an isomorphism $u : \mathbb{R}^k \to E_x$, which can also be viewed as a basis $(v_1, \ldots, v_k)$ of E_x (the image by u of the canonical basis of $\mathbb{R}^k$).

For every $x \in M$ and tangent vector $X \in T_x M$, we can define the horizontal lift of X to any $u = (v_1, \ldots, v_k) \in \mathrm{Gl}_x(E)$ in the following way. Take local sections σ_i of E around x such that $\sigma_i(x) = v_i$ and $(\nabla_X \sigma_i)_x = 0$. Then $\sigma := (\sigma_1, \ldots, \sigma_k)$ is a local section of $\mathrm{Gl}(E)$ satisfying $\sigma(x) = u$. We define $(\tilde{X})_u := \sigma_*(X) \in T_u \mathrm{Gl}(E)$. It can be easily shown, using a local trivialization like in Lemma 5.2, that $\tilde{X}$ does not depend on the local sections σ_i. If π denotes the canonical projection $\pi : \mathrm{Gl}(E) \to M$, we have by construction $\pi_*(\tilde{X}_u) = X$ for all $u \in \mathrm{Gl}_x(E)$. The set of all horizontal lifts at u is an n-dimensional subspace of $T_u \mathrm{Gl}(E)$, called the horizontal subspace and denoted by $T_u^h \mathrm{Gl}(E)$, or simply $\mathcal{H}_u$ when no confusion is possible. The collection of all horizontal spaces is called the horizontal distribution. If we denote by $\mathcal{V}$ the vertical distribution on $\mathrm{Gl}(E)$ consisting in the tangent spaces to the fibres, $\mathcal{V}_u := T_u(\mathrm{Gl}_{\pi(u)}(E)) \subset T_u(\mathrm{Gl}(E))$, then we have as before a direct sum decomposition $T\mathrm{Gl}(E) = \mathcal{H} \oplus \mathcal{V}$. Notice that $\mathcal{V}$ is always defined independently of ∇, while $\mathcal{H}$ depends heavily on the covariant derivative.

We claim that the horizontal distribution $\mathcal{H}$ is invariant under the right action of $\mathrm{Gl}_k(\mathbb{R})$ on $\mathrm{Gl}(E)$, as a consequence of the linearity of ∇. Let $a = (a_{ij}) \in \mathrm{Gl}_k(\mathbb{R})$, $X \in T_x M$, $u = (v_1, \ldots, v_k) \in \mathrm{Gl}_x(E)$ and let $\sigma := (\sigma_1, \ldots, \sigma_k)$ be a local section of $\mathrm{Gl}(E)$ satisfying $\sigma(x) = u$ and $(\nabla_X \sigma_i)_x = 0$. Then σa is a local section of $\mathrm{Gl}(E)$ with $(\sigma a)(x) = ua$ and its components $(\sigma a)_i = \sum_j a_{ji} \sigma_j$ satisfy $\nabla_X (\sigma a)_i = 0$ at x because ∇ is $\mathbb{R}$-linear. We thus have

$$(R_a)_*(\tilde{X}_u) = (R_a)_*(\sigma_*(X)) = (R_\alpha \circ \sigma)_*(X) = (\sigma a)_*(X) = \tilde{X}_{ua}.$$

This motivates the following:

DEFINITION 5.5. *Let P be a G-principal bundle over a manifold M with vertical distribution (tangent to the fibres) denoted by $\mathcal{V}$. A connection on P is a smooth distribution $\mathcal{H}$ such that $TP = \mathcal{H} \oplus \mathcal{V}$ and $(R_a)_*(\mathcal{H}_u) = \mathcal{H}_{ua}$ for all $u \in P$ and $a \in G$.*

The first condition says that π_* maps $\mathcal{H}_u$ isomorphically over $T_{\pi(u)}M$ for every $u \in P$. The inverse image through this map of a vector $X \in T_{\pi(u)}M$ is called the *horizontal lift* of X at u.

DEFINITION 5.6. *If x is some point of M, a local section σ of P in a neighbourhood of x is called* horizontal at x *if its differential $d\sigma_x$ at x maps T_xM into $\mathcal{H}_{\sigma(x)}$.*

LEMMA 5.7. *For every $x \in M$ and $u \in P$ with $\pi(u) = x$ there exist local sections of P horizontal at x.*

PROOF. There exist local coordinates near u and x, in which $u = 0$ and the submersion π can be written as a linear map $\pi(x_1, \ldots, x_N) = (x_1, \ldots, x_n)$. The section of P which corresponds in these coordinates to the inverse of the isomorphism $\pi_* : \mathcal{H}_u \to T_xM$ is then horizontal at x. $\square$

THEOREM 5.8. *A covariant derivative on a vector bundle E over a manifold M induces a connection on its frame bundle $\mathrm{Gl}(E)$. Conversely, a connection on a G-principal bundle P over M induces in a canonical way a covariant derivative on all vector bundles associated to P by a linear representation ρ of G. If $G = \mathrm{Gl}_k(\mathbb{R})$ and ρ is the identity, these two constructions are inverse to each other.*

PROOF. The direct statement was proved above. Conversely, let $\rho : G \to \mathrm{Gl}_k(\mathbb{R})$ be some representation of G and let $E := P \times_\rho \mathbb{R}^k$ be the associated vector bundle. Given a local section σ of P over some open set $U \subset M$, any section ψ of E over U can be expressed in the frame σ as an $\mathbb{R}^k$-valued function ξ on U: $\psi = [\sigma, \xi]$. The main idea is to choose a suitable local frame σ on P using the horizontal distribution and to define the covariant derivative of the local section ψ as the section which in the frame σ corresponds to the usual directional derivative of the function ξ.

More precisely, if x is a fixed point of M, any local section ψ of E can be written $\psi = [\sigma, \xi]$ for some smooth function ξ defined around x with values in $\mathbb{R}^k$ and some local section σ horizontal at x. For $X \in T_xM$ we define

$$(\nabla_X \psi)_x := [\sigma(x), \partial_X(\xi)]. \qquad (5.2)$$

Let us first show that this formula does not depend on the choice of σ. If σ' is another local section of P horizontal at x, we can write $\sigma' = \sigma f$ for some local map f defined around x with values in G. We denote $f(x)$ by a and let $A \in T_a G$ be the image by f_* of some $X \in T_x M$. The Leibniz rule yields

$$\sigma'_*(X) = (R_a)_*(\sigma_*(X)) + (L_u)_*(A).$$

By assumption $\sigma'_*(X)$ and $\sigma_*(X)$ both belong to $\mathcal{H}$, and since $\mathcal{H}$ is G-invariant, we obtain $(L_u)_*(A) \in \mathcal{H}$. On the other hand $(L_u)_*(A) \in \mathcal{V}$ by definition, so $(L_u)_*(A) = 0$. Since L_u is a diffeomorphism, we must have $A = 0$. The expression of the section ψ in the frame σ' is $\psi = [\sigma', \rho(f^{-1})\xi]$, so using the Leibniz rule again

$$[\sigma'(x), \partial_X(\rho(f^{-1})\xi)] = [\sigma'(x), \rho(f^{-1})(\partial_X \xi)] - [\sigma'(x), \rho_*(A)\xi] = [\sigma(x), \partial_X \xi].$$

This shows that ∇ is well-defined. The fact that ∇ satisfies the axioms of a covariant derivative is an easy exercise. $\qquad \square$

5.3. Linear connections

Let M be a smooth manifold.

DEFINITION 5.9. *A* linear connection *on M is a connection in the frame bundle of M or, equivalently, a covariant derivative on TM.*

LEMMA 5.10. *Let ∇ be the covariant derivative of a linear connection on M. Then the expression*

$$T^\nabla(X, Y) := \nabla_X Y - \nabla_Y X - [X, Y], \qquad \forall\, X, Y \in \mathcal{X}(M)$$

defines a tensor field of type $(2, 1)$ called the torsion *of ∇.*

PROOF. Let X, Y be vector fields and let f and h be two arbitrary functions on M. Using (2.8) and Definition 5.1 we compute

$$
\begin{aligned}
T^\nabla(fX, hY) &= \nabla_{fX}(hY) - \nabla_{hY}(fX) - [fX, hY] \\
&= fh\nabla_X Y + f(\partial_X h)X - fh\nabla_Y X - h(\partial_Y f)X \\
&\quad - (fh[X, Y] + f(\partial_X h)Y - h(\partial_Y f)X) \\
&= fhT^\nabla(X, Y).
\end{aligned}
$$

Thus T^∇ is $C^\infty(M)$-bilinear, so it defines a tensor field by Proposition 2.3. $\square$

A linear connection is called *torsion-free* if its torsion vanishes.

5.4. Pull-back of bundles

Let $f : M \to N$ be a smooth map and let $\pi : P \to N$ be a G-principal bundle over N. The *pull-back* of P by f is

$$f^* P := \{(u, x) \in P \times M \mid \pi(u) = f(x)\}.$$

From the local triviality of P we see that the map $f^* P \to M$ given by $(u, x) \mapsto x$ is a G-principal bundle over M. The fibre $(f^* P)_x$ is canonically identified with the fibre $P_{f(x)}$ by the map $(u, x) \mapsto u$. Since f^*P is a subset of $P \times M$, its tangent space at some (u, x) is a subset of $T_uP \times T_xM$. More precisely,

$$T_{(u,x)}f^*P = \{(V, X) \in T_uP \times T_xM \mid \pi_*(V) = f_*(X)\}.$$

The notion of pull-back can be transposed verbatim to vector bundles, the details being left to the reader. The following lemma is an easy play with the definitions:

LEMMA 5.11. *Let $f : M \to N$ be some smooth map. If $E \to N$ is a vector bundle, then $\mathrm{Gl}(f^*E) = f^*(\mathrm{Gl}(E))$. Similarly, if $P \to N$ is a G-structure and ρ is a representation of G on some vector space V, then $f^*(P \times_\rho V) = (f^*P) \times_\rho V$.*

A connection $\mathcal{H}$ on P induces canonically a pull-back connection $f^*\mathcal{H}$ on f^*P by the formula

$$(f^*\mathcal{H})_{(u,x)} := \{(V, X) \in \mathcal{H}_u \times T_xM \mid \pi_*(V) = f_*(X)\}. \tag{5.3}$$

Correspondingly, a covariant derivative ∇ on a vector bundle E induces a pull-back covariant derivative on any pull-back f^*E: if $\mathcal{H}$ denotes the connection on $\mathrm{Gl}(E)$ corresponding to ∇ given by Theorem 5.8, we define $f^*\nabla$ to be the covariant derivative on f^*E which induces the connection $f^*\mathcal{H}$ on $f^*\mathrm{Gl}(E) = \mathrm{Gl}(f^*E)$.

5.5. Parallel transport

Let $\pi : P \to M$ be a G-structure with a connection $\mathcal{H}$ on it. A path u_t in P is called *horizontal* if $\dot{u}_t \in \mathcal{H}_{u_t}$ for all t.

PROPOSITION 5.12. *For every smooth path $x : [0, 1] \to M$, $t \mapsto x_t$ and $u \in P$ such that $\pi(u) = x_0$, there exists a unique horizontal path u_t in P such that $\pi(u_t) = x_t$ for all $t \in [0, 1]$. The path u_t is called the* horizontal lift *of x_t through u.*

PROOF. The fact that x is smooth means that there exists a smooth path defined on some open interval I containing $[0, 1]$ whose restriction to $[0, 1]$ is x. We first prove a local version of the statement.

On the pull-back bundle x^*P (which is a G-structure over I) let X denote the horizontal lift of the standard vector field $\partial/\partial t \in \mathcal{X}(I)$ with respect to the pull-back connection. From Proposition 1.11, for every $t_0 \in I$ and $u \in \pi^{-1}(x_{t_0})$, there exists a unique integral curve (t, u_t) of X in x^*P defined on some open interval U_{t_0} containing t_0. By definition $\pi(u_t) = x(t) = x_t$, so u_t is a local lift of x_t through u, and (5.3) shows that u_t is horizontal in P.

In order to prove the global existence of the horizontal lift, we notice that the invariance of the connection insures that if u_t is a horizontal path, then

$u_t a$ is also horizontal for every $a \in G$. Consequently, the open set of definition U_{t_0} of the local horizontal lifts defined above does not depend on the element u of the fibre $\pi^{-1}(x_{t_0})$. By compactness, we can choose a finite number of such neighbourhoods which cover $[0, 1]$, and construct the global horizontal lift inductively, using again the right invariance. The uniqueness is clear. □

The element u_1 is called the *parallel transport* of u_0 along the path x_t. This notion obviously makes sense if one assumes that x_t is only piecewise smooth. In general the parallel transport depends on the whole path x_t, not only on its endpoint.

5.6. Holonomy

The dependence of the parallel transport on the path used to define it gives rise to the notion of *holonomy*. Let $\pi : P \to M$ be a G-structure with a connection on it. The *holonomy group* $\mathrm{Hol}(u)$ at some $u \in P$ is defined to be the set of all $a \in G$ such that u and ua can be joined by a piecewise smooth horizontal path. Similarly, we define the *restricted holonomy group* $\mathrm{Hol}_0(u)$ as the set of all $a \in G$ such that u and ua can be joined by a piecewise smooth horizontal path whose projection to M is contractible. Of course, $\mathrm{Hol}(u) = \mathrm{Hol}_0(u)$ if M is simply connected.

LEMMA 5.13. (*i*) *The holonomy group* $\mathrm{Hol}(u)$ *is a subgroup of* G.

(*ii*) $\mathrm{Hol}(ua) = a^{-1}\mathrm{Hol}(u)a$ *for every* $u \in P$ *and* $a \in G$.

(*iii*) *If* u *and* v *can be joined by a horizontal path, then* $\mathrm{Hol}(u) = \mathrm{Hol}(v)$.

PROOF. (*i*) Clearly $e \in \mathrm{Hol}(u)$. If $a, b \in \mathrm{Hol}(u)$, let γ and γ' be two horizontal paths in P joining u with ua and ub respectively. By right invariance, the composition $\gamma' \cdot (\gamma b)$ is a piecewise smooth horizontal path joining u with uab. Similarly, $\gamma^{-1}a^{-1}$ joins u with ua^{-1}, where $\gamma^{-1}(t)$ denotes the path $\gamma(1-t)$.

(*ii*) For every $b \in \mathrm{Hol}(u)$ there exists a horizontal path γ joining u with ub. The horizontal path γa then joins ua with $uba = ua(a^{-1}ba)$, thus showing that $a^{-1}\mathrm{Hol}(u)a \subset \mathrm{Hol}(ua)$. The other inclusion is similar.

(*iii*) We denote by β the horizontal path joining u with v. For $a \in \mathrm{Hol}(u)$, let γ be a horizontal path joining u with ua. Then $\beta^{-1} \cdot \gamma \cdot \beta a$ joins v with va, so $a \in \mathrm{Hol}(v)$. Thus $\mathrm{Hol}(u) \subset \mathrm{Hol}(v)$, so by symmetry, the two holonomy groups are equal. □

The holonomy group $\mathrm{Hol}(u)$ is actually a Lie subgroup of G, and $\mathrm{Hol}_0(u)$ is the connected component of the identity element in $\mathrm{Hol}(u)$ (cf. [10], Theorem 4.2, p. 73).

If M is connected, for every $u, v \in P$ one can choose a smooth path c on M joining $\pi(u)$ with $\pi(v)$. The horizontal lift of c through u then joins u with some element of the fibre of v, so by the previous lemma, the holonomy groups $\mathrm{Hol}(u)$ and $\mathrm{Hol}(v)$ are conjugate. By a slight abuse of language, we define the holonomy group of M, denoted $\mathrm{Hol}(M)$, as the conjugacy class in G of the holonomy subgroups $\mathrm{Hol}(u)$, $u \in P$.

Let $P(u)$ denote the set of $v \in P$ which can be joined with u by a piecewise smooth horizontal path. By Lemma 5.13, the restriction of the right action of G to $\mathrm{Hol}(u)$ preserves $P(u)$, which thus defines a reduction of the structure group of P to $\mathrm{Hol}(u)$. The fact that $P(u)$ is a smooth submanifold of P follows from the local existence of sections of P taking values in $P(u)$, which is itself a consequence of the integrability theorem of Frobenius (see Section 2.7 in [10] for details). The principal bundle $P(u)$ is called the *holonomy bundle* through u.

5.7. Reduction of connections

Let $Q \subset P$ be a reduction of a G-structure P to a subgroup H of G. A connection $\mathcal{H}$ on P is called *reducible* to Q if the horizontal space $\mathcal{H}_u$ is a subspace of $T_u Q$ for all $u \in Q$. If $\mathcal{H}$ is reducible to Q, it clearly defines a connection $\mathcal{H}'$ on Q by $\mathcal{H}'_u := \mathcal{H}_u$ for all $u \in Q$.

LEMMA 5.14. *A connection $\mathcal{H}$ on P is reducible to Q if and only if for every $u_0 \in Q$ and for every curve x_t in M with $x_0 = \pi(u_0)$, the parallel transport of u_0 along x_t belongs to Q.*

PROOF. Let u_t denote the horizontal lift of x_t through u_0 with respect to $\mathcal{H}$. If $\mathcal{H}$ is reducible to a connection $\mathcal{H}'$ on Q, let u'_t denote the horizontal lift of x_t through u_0 with respect to $\mathcal{H}'$. Since $\dot{u}'_t \in \mathcal{H}'_{u'_t} = \mathcal{H}_{u'_t}$, we see that u'_t is also a horizontal lift of x_t with respect to $\mathcal{H}$, so $u'_t = u_t$ by Proposition 5.12, and thus $u_t \in Q$.

Conversely, let $Y_{u_0} \in \mathcal{H}_{u_0}$ be any horizontal tangent vector. Consider a path x_t in M such that $x_0 = \pi(u_0)$ and $\dot{x}_0 = \pi_*(Y)$. Since the horizontal lift u_t of x_t through u_0 belongs to Q we deduce that $Y_{u_0} = \dot{u}_0 \in T_{u_0}Q$, thus showing that $\mathcal{H}_{u_0}$ is a subspace of $T_{u_0}Q$. $\square$

By the definition of holonomy bundles we thus have the following:

COROLLARY 5.15. *A connection on a principal bundle P is reducible to all its holonomy bundles $P(u)$.*

Our aim is to understand under which circumstances a connection reduces to a subbundle defined by a section in some associated vector bundle to P as in Proposition 4.7.

Let P be a G-structure over M, and let $E = P \times_\rho V$ be the associated vector bundle to some representation ρ of G on a vector space V. If H denotes the stabilizer of some element of V

$$H := \{g \in G \mid \rho(g)\xi = \xi\}$$

and $\mathcal{O}$ is the G-orbit of ξ in V, the associated fibration $P \times_\rho \mathcal{O}$ is a subset of E, which we previously denoted by E^ξ. Recall that by Proposition 4.7, sections σ of E^ξ are in one-to-one correspondence with H-subbundles of P.

THEOREM 5.16. *Let σ be a section of $E^\xi \subset E$ corresponding to an H-subbundle Q of P. If $\mathcal{H}$ is a connection on P and ∇ denotes the corresponding covariant derivative induced on E, then the connection $\mathcal{H}$ is reducible to Q if and only if $\nabla \sigma = 0$.*

PROOF. Let ξ be some fixed element of V. By definition, the distinguished section σ of E^ξ can be written $\sigma = [u, \xi]$ where u is an arbitrary section of Q.

If $\mathcal{H}$ is reducible to a connection $\mathcal{H}'$ on Q, we can always choose u to be horizontal with respect to $\mathcal{H}'$ at some point x, so (5.2) shows that $\nabla \sigma = 0$ at x.

Conversely, if $\nabla \sigma = 0$, let u_0 be some arbitrary element of Q and let x_t be some path in M with $x_0 = \pi(u_0)$. We denote by u_t the horizontal lift of x_t through u_0. Then σ and $\sigma' := [u_t, \xi]$ are two parallel sections along x_t which coincide at x_0. By Lemma 5.4 we have $\sigma = \sigma'$ along x_t, so $u_t \in Q$, by the very definition (4.2) of Q. Lemma 5.14 thus shows that $\mathcal{H}$ is reducible to Q. $\square$

5.8. Exercises

(1) If $\mathcal{H}$ is a connection on a principal bundle P, prove the existence of local sections which are horizontal at a point $x \in M$. If σ is such a section and $X \in T_x M$, prove that $\sigma_*(X)$ is the horizontal lift of X at $\sigma(x)$.

(2) Check the last statement of Theorem 5.8.

(3) Prove Lemma 5.11.

(4) Prove that formula (5.3) defines a connection on the pull-back of P.

(5) (*The connection form*) If $\mathcal{H}$ is a connection in a G-principal bundle $\pi : P \to M$, let $\omega^{\mathcal{H}}$ be the $\mathfrak{g}$-valued 1-form on P defined for every $u \in P$ and $U \in T_u P$ by

$$\omega_u^{\mathcal{H}}(U) := (L_u)_*^{-1}(\pi_{\mathcal{V}_u} U),$$

where $\pi_{\mathcal{V}_u}$ denotes the projection onto $\mathcal{V}_u$ parallel to H_u and $(L_u)_*$ is as usual the isomorphism between $\mathfrak{g}$ and $\mathcal{V}_u$ given by the differential at the origin of the map $L_u : G \to P$, $a \mapsto ua$. Use the right invariance of $\mathcal{H}$ to prove the following equivariance property of $\omega^{\mathcal{H}}$:

$$(R_a)^* \omega^{\mathcal{H}} = (\mathrm{ad}_{a^{-1}})_* \circ \omega^{\mathcal{H}}, \tag{5.4}$$

where $\mathrm{ad}_{a^{-1}} : G \to G$ denotes the adjoint morphism $g \mapsto a^{-1}ga$.

Conversely, show that every $\mathfrak{g}$-valued 1-form ω on P satisfying $\omega(U) = (L_u)_*^{-1}(U)$ for all $U \in \mathcal{V}_u$ and (5.4) defines a connection $\mathcal{H}^\omega$ on P by the formula $\mathcal{H}^\omega := \ker \omega$.

CHAPTER 6

Riemannian manifolds

6.1. Riemannian metrics

Let M be a smooth manifold.

DEFINITION 6.1. *A Riemannian metric on M is tensor field g of type $(2,0)$ which (as a bilinear form) is symmetric and positive definite at each point of M.*

By Proposition 4.7, a Riemannian metric g on a smooth n-dimensional manifold M, is equivalent to a reduction $O(M)$ of the frame bundle $Gl(M)$ to the group $O(n)$, where $O(M)$ consists exactly in those frames which are orthonormal with respect to g.

A pair (M, g) where g is a Riemannian metric on M is called a *Riemannian manifold*. Each tangent space $T_x M$ of a Riemannian manifold is thus a Euclidean space with respect to the bilinear form g_x. Every vector $\xi \in T_x M$ on a Riemannian manifold defines a 1-form $\xi^\flat$ by the formula

$$\xi^\flat(X) := g(\xi, X), \qquad \forall\, X \in T_x M.$$

Similarly, an endomorphism A of $T_x M$ defines a bilinear form $A^\flat$ by the formula

$$A^\flat(X, Y) := g(A(X), Y), \qquad \forall\, X, Y \in T_x M.$$

When there is no risk of confusion (e.g., when the Riemannian metric is fixed), we will sometimes identify a vector (or an endomorphism) and the corresponding 1-form (resp. bilinear form).

An important observation is that on Riemannian manifolds one can measure the length of curves. If $\gamma : [0, 1] \to M$ is a smooth simple curve, its length is defined to be

$$l(\gamma) := \int_0^1 g(\dot\gamma(t), \dot\gamma(t))^{1/2} dt,$$

and this definition depends only on the curve itself, not on its parametrization. Indeed, if $t = t(s)$ is a smooth direct diffeomorphism of the interval $[0, 1]$ and $c(s) := \gamma(t(s))$ then $\dot c(s) = \dot\gamma(t(s)) \cdot t'(s)$ so $g(\dot c(s), \dot c(s))^{1/2} = g(\dot\gamma(t), \dot\gamma(t))^{1/2} \cdot t'(s)$ and the claim follows from the classical change of variable formula.

Riemannian manifolds are metric spaces with respect to the distance defined by
$$d(x, y) := \inf\{l(\gamma) \mid \gamma(0) = x, \gamma(1) = y\},$$
but we will not further develop this point of view here.

Every smooth manifold M^n carries Riemannian metrics. To see this, take an atlas (U_i, ϕ_i) of M and a partition of unity (f_i) subordinate to the open cover $\{U_i\}$. For every $x \in U_i$, let $(g_i)_x$ denote the pull-back of the standard Riemannian metric on $\mathbb{R}^n$ by the map ϕ_i. Then $g := \sum f_i g_i$ is a well-defined Riemannian metric on M.

6.2. The Levi–Civita connection

The most important feature of Riemannian manifolds is the fact that they carry a distinguished linear connection.

PROPOSITION 6.2. *On any Riemannian manifold (M, g) there exists a unique torsion-free linear connection which is reducible to the bundle of orthonormal frames $O(M)$.*

PROOF. For every linear connection ∇, we will denote by the same symbol the covariant derivative induced by ∇ on the tensor bundle which satisfies the Leibniz rule with respect to the tensor product, commutes with contractions and equals the usual vector derivative on functions.

From Theorem 5.16, we need to show that there exists a unique linear connection ∇ such that $T^{\nabla} = 0$ and $\nabla g = 0$. The first relation reads
$$\nabla_X Y - \nabla_Y X = [X, Y], \qquad \forall\, X, Y \in \mathcal{X}(M). \tag{6.1}$$
In order to exploit the second relation, we notice that if X, Y and Z are vector fields on M, we can write
$$\begin{aligned}
\partial_X(g(Y, Z)) &= \nabla_X(g(Y, Z)) = \nabla_X(C_1 C_2(g \otimes Y \otimes Z)) \\
&= C_1 C_2(\nabla_X(g \otimes Y \otimes Z)) \\
&= (\nabla_X g)(Y, Z) + g(\nabla_X Y, Z) + g(Y, \nabla_X Z).
\end{aligned}$$
Consequently, $\nabla g = 0$ is equivalent to
$$\partial_X(g(Y, Z)) = g(\nabla_X Y, Z) + g(Y, \nabla_X Z), \qquad \forall\, X, Y, Z \in \mathcal{X}(M). \tag{6.2}$$
We perform circular permutations in this equation and use (6.1) to obtain
$$\begin{aligned}
\partial_X(g(Y, Z)) + \partial_Y(g(Z, X)) - \partial_Z(g(X, Y)) &= g(\nabla_X Y, Z) + g(Z, \nabla_Y X) \\
&\quad + g(Y, [X, Z]) + g(X, [Y, Z])
\end{aligned}$$
so by (6.1) again
$$\begin{aligned}
2g(\nabla_X Y, Z) &= \partial_X(g(Y, Z)) + \partial_Y(g(Z, X)) - \partial_Z(g(X, Y)) \\
&\quad + g(Y, [Z, X]) + g(Z, [X, Y]) - g(X, [Y, Z]).
\end{aligned} \tag{6.3}$$

Since g is non-degenerate, this shows the uniqueness of ∇. Conversely, this formula defines an operator ∇ which is easily seen to satisfy the axioms of Definition 5.1, as well as (6.1) and (6.2). $\qquad\square$

The connection given by the proposition above is called the *Levi–Civita connection*. From now on, if (M, g) is a Riemannian manifold, we will denote by ∇ the Levi–Civita covariant derivative of g, without stating this explicitly each time. A Riemannian manifold (M^n, g) is called *(locally) irreducible* if the (restricted) holonomy groups of the Levi–Civita connection (recall that they are all conjugate) act irreducibly on $\mathbb{R}^n$.

A special case of Lemma 5.7 which turns out to be very useful for concrete computations is the following:

LEMMA 6.3. *Around every point x in a Riemannian manifold (M^n, g) there exists a local orthonormal frame $u = \{e_1, \ldots, e_n\}$ parallel at x with respect to ∇.*

Let (M^n, g) be an oriented Riemannian manifold. Since the representation of $\mathrm{Gl}_n(\mathbb{R})$ on $\Lambda^n \mathbb{R}^n$ is given by the determinant, its restriction to SO_n is trivial, thus showing that the vector bundle $\Lambda^n M$ has a distinguished section $dv := [u, \zeta]$ where u is any oriented orthonormal frame on M and $\zeta \in \Lambda^n \mathbb{R}^n$ is the canonical unit element. The n-form dv is called the *volume form*. If $\{\varepsilon_i\}$ is an oriented local orthonormal coframe, then $dv = \varepsilon_1 \wedge \cdots \wedge \varepsilon_n$.

We end this section by a result relating the Lie derivative and the covariant derivative.

LEMMA 6.4. *If ξ is a vector field on a Riemannian manifold (M, g), the Lie derivative of g with respect to ξ equals twice the symmetric part of $\nabla \xi$.*

PROOF. Easy computation:

$$
\begin{aligned}
(\mathcal{L}_\xi g)(X, Y) \quad &= \quad \mathcal{L}_\xi(g(X, Y)) - g(\mathcal{L}_\xi X, Y) - g(X, \mathcal{L}_\xi Y) \\
&= \quad \partial_\xi(g(X, Y)) - g([\xi, X], Y) - g(X, [\xi, Y]) \\
&\overset{(6.1),(6.2)}{=} \quad g(\nabla_\xi X, Y) + g(X, \nabla_\xi Y) - g(\nabla_\xi X - \nabla_X \xi, Y) \\
&\qquad -g(X, \nabla_\xi Y - \nabla_Y \xi) \\
&= \quad g(\nabla_X \xi, Y) + g(X, \nabla_Y \xi).
\end{aligned}
$$

$\qquad\square$

6.3. The curvature tensor

To any covariant derivative on a vector bundle E one can associate its *curvature* which is a $(2, 0)$-tensor with values in the bundle $\mathrm{End}(E) \cong E \otimes E^*$. This point of view will be developed later on (in Section 10.1). For the moment, we consider the particular case of a linear connection ∇ on a manifold M.

LEMMA 6.5. *The operator*

$$R^\nabla(X,Y)Z := \nabla_X \nabla_Y Z - \nabla_Y \nabla_X Z - \nabla_{[X,Y]}Z, \quad \forall\, X,Y,Z \in \mathcal{X}(M)$$

defines a tensor of type $(3,1)$ *called the* curvature tensor *of* ∇.

PROOF. By Proposition 2.3, this follows from the $C^\infty(M)$-multilinearity of R^∇, which is straightforward to check using Definition 5.1. $\qquad\square$

If (M^n, g, ∇) is a Riemannian manifold, it is sometimes convenient to view the curvature tensor as a $(4,0)$-tensor by:

$$R(X,Y,Z,T) := g(R^\nabla(X,Y)Z,T), \quad \forall\, X,Y,Z,T \in TM,$$

called the *Riemannian curvature tensor.*

A Riemannian metric is called *flat* if its Riemannian curvature tensor vanishes. Using the integrability theorem of Frobenius, one can show that this is equivalent to the fact that M is locally isometric to a Euclidean space.

The following classical result can be checked by direct calculation but its real signification can only be understood in the context of connection forms (see Exercise (5) in Section 5.8). Details can be found in [10], p. 135.

LEMMA 6.6. *The fully covariant Riemannian curvature tensor R of a Riemannian manifold has the following symmetries:*

- $R(X,Y,Z,T) = -R(X,Y,T,Z)$;
- $R(X,Y,Z,T) = R(Z,T,X,Y)$;
- $R(X,Y,Z,T) + R(Y,Z,X,T) + R(Z,X,Y,T) = 0$
  (*1st Bianchi identity*);
- $(\nabla_X R)(Y,Z,T,W) + (\nabla_Y R)(Z,X,T,W) + (\nabla_Z R)(X,Y,T,W) = 0$
  (*2nd Bianchi identity*).

The *Ricci tensor* of (M,g) is defined by

$$\mathrm{Ric}(X,Y) := \mathrm{Tr}\{V \mapsto R(V,X)Y\},$$

or equivalently

$$\mathrm{Ric}(X,Y) = \sum_{i=1}^{2m} R(e_i, X, Y, e_i),$$

where e_i is a local orthonormal basis of TM. We recall that the Ricci tensor of every Riemannian manifold is symmetric, as can be easily seen from the symmetries of the Riemannian curvature. A Riemannian metric g on a manifold M is called *Einstein* if its Ricci tensor Ric is proportional to the metric tensor, that is, if there exists a real constant λ such that

$$\mathrm{Ric}(X,Y) = \lambda g(X,Y), \quad \forall\, X,Y \in TM.$$

If $\lambda = 0$, the metric g is called *Ricci-flat*. Using Lemma 6.6 it is easy to show that a Ricci-flat metric in dimension 2 is flat. With a little more work one can show the following:

PROPOSITION 6.7. *A compact manifold of dimension 2 admitting a flat Riemannian metric is isometric to a flat torus $\mathbb{R}^2/\Gamma$ where Γ is a group generated by two non-collinear translations.*

The existence of Einstein metrics on a given manifold is one of the major open problems in modern Riemannian geometry. As we will see later on, this problem can be partially solved in the framework of Kähler geometry.

6.4. Killing vector fields

Let (M, g) be a Riemannian manifold. An *isometry* of M is a diffeomorphism $f : M \to M$ such that $f^*g = g$, i.e.

$$g(f_*(X), f_*(Y))_{f(x)} = g(X, Y)_x, \qquad \forall\, x \in M, \, \forall\, X, Y \in T_x M.$$

DEFINITION 6.8. *A vector field on a Riemannian manifold is called a Killing vector field or* infinitesimal isometry *if its local flow consists of (local) isometries of M.*

This definition has a nice geometrical meaning, but in practice one often uses the following alternative characterization:

PROPOSITION 6.9. *Let ξ be a vector field on a Riemannian manifold (M, g, ∇). The following statements are equivalent:*

(i) ξ is Killing with respect to g.

(ii) The Lie derivative of g with respect to ξ vanishes: $\mathcal{L}_\xi g = 0$.

(iii) The covariant derivative $\nabla \xi$ is skew-symmetric with respect to g:

$$g(\nabla_X \xi, Y) + g(\nabla_Y \xi, X) = 0, \qquad \forall\, X, Y \in TM. \tag{6.4}$$

PROOF. Let φ_t denote the local flow of ξ. If ξ is Killing, φ_t are local isometries so by (2.6)

$$\mathcal{L}_\xi g = -\frac{d}{dt}\Big|_{t=0} (\varphi_t)_*(g) = \frac{d}{dt}\Big|_{t=0} (\varphi_t)^*(g) = \frac{d}{dt}\Big|_{t=0} g = 0.$$

Conversely, suppose that $\mathcal{L}_\xi g = 0$. Since $(\varphi_t)_* \xi = \xi$, the naturality of the Lie derivative (2.7) yields

$$0 = (\varphi_s)^*(\mathcal{L}_\xi g) = \mathcal{L}_\xi((\varphi_s)^*g)$$

for all fixed s. Taking into account the fact that φ_t is a pseudogroup of local diffeomorphisms (1.7) we get

$$0 = \mathcal{L}_\xi((\varphi_s)^*g) = \frac{d}{dt}\Big|_{t=0}(\varphi_t)^*((\varphi_s)^*g) = \frac{d}{dt}\Big|_{t=0}(\varphi_{s+t})^*(g)$$
$$= \frac{d}{dt}\Big|_{t=s}(\varphi_t)^*g.$$

This shows that the family of tensors $(\varphi_t)^*g$ does not depend on t, so $(\varphi_t)^*g = (\varphi_0)^*g = g$, thus proving the equivalence of (i) and (ii).

The equivalence of (ii) and (iii) is a consequence of Lemma 6.4.   □

REMARK. The isometry group of every compact manifold is a Lie group, whose Lie algebra is isomorphic to the space of Killing vector fields.

6.5. Exercises

(1) Show that the holonomy group $\mathrm{Hol}(u)$ at some orthonormal frame u of a Riemannian manifold (M^n, g) is contained in the orthogonal group O_n.

(2) (*Conformal metric changes*) Let (M, g) be a Riemannian manifold and let $u : M \to \mathbb{R}$ be any function. We define a new Riemannian metric on M by $\bar{g} = e^{2u}g$. Show that the Levi–Civita covariant derivatives ∇ and $\bar{\nabla}$ of g and $\bar{g}$ are related by

$$\bar{\nabla}_X Y = \nabla_X Y + (\partial_X u)Y + (\partial_Y u)X - g(X,Y)\mathrm{Grad}_g u \ ,$$

where $\mathrm{Grad}_g u$ is the dual of du with respect to g.

(3) Let (M, g, ∇) be a Riemannian manifold. If η denotes a 1-form on M, we may view its covariant derivative $\nabla\eta$ as a bilinear form on TM by the formula $\nabla\eta(X,Y) := (\nabla_X\eta)(Y)$. Show that the exterior derivative of η equals twice the skew-symmetric part of $\nabla\eta$:

$$d\eta(X,Y) = \nabla\eta(X,Y) - \nabla\eta(Y,X).$$

(4) Prove that the set of Killing vector fields is a Lie algebra with respect to the usual Lie bracket. *Hint:* Use (2.9) and Proposition 6.9.

(5) (*Kostant formula*) Let ξ be a Killing vector field on a Riemannian manifold (M, g, ∇) with Riemannian curvature tensor R. Prove that the covariant derivative of the (1,1)-tensor $\nabla\xi$ satisfies

$$\nabla_X(\nabla\xi) = R(X,\xi), \qquad \forall\, X \in TM.$$

Hint: If X, Y, Z are vector fields on M, take the covariant derivative of (6.4) with respect to Z and use circular permutations in X, Y, Z.

(6) Show that a Killing vector field ξ is determined by its 1-jet at any point. *Hint:* Use the previous exercise to show that $(\xi, \nabla\xi)$ is a parallel section of the bundle $TM \oplus \mathrm{End}(TM)$ with respect to a certain covariant derivative on that bundle, and apply Lemma 5.4 to conclude that if ξ and $\nabla\xi$ both vanish at some point, then ξ is identically zero.

(7) Show that the space of Killing vector fields on a Riemannian manifold of dimension n is at most $\frac{n(n+1)}{2}$-dimensional.

(8) Show that the volume form of an oriented Riemannian manifold is parallel with respect to the Levi–Civita covariant derivative.

(9) Let ∇ be a torsion-free covariant derivative on the tangent bundle of a smooth manifold M. Using (3.6), show that the exterior derivative is related to ∇ by the following formula:

$$d\omega(X_0, \dots, X_p) = \sum_{i=0}^{p} (-1)^i (\nabla_{X_i}\omega)(X_0, \dots, \widehat{X}_i, \dots, X_p), \qquad (6.5)$$

for all $X_0, \dots, X_p \in \mathcal{X}(M)$ and $\omega \in \Omega^p M$.

Part 2

Complex and Hermitian geometry

CHAPTER 7

Complex structures and holomorphic maps

7.1. Preliminaries

Kähler manifolds may be considered as special Riemannian manifolds. Besides the Riemannian structure, they also have compatible symplectic and complex structures.

DEFINITION 7.1. *A Kähler structure on a Riemannian manifold* (M^n, g) *is given by a 2-form* Ω, *called Kähler form and a field of endomorphisms* J *of the tangent bundle, satisfying the following*

- algebraic conditions
  - (a) *J is an almost complex structure: $J^2 = -Id$.*
  - (b) *The metric is almost Hermitian with respect to J:* $g(X, Y) = g(JX, JY)$, $\forall X, Y \in TM$.
  - (c) $\Omega(X, Y) = g(JX, Y)$.
- analytic conditions
  - (d) *the 2-form Ω is closed: $d\Omega = 0$.*
  - (e) *J is integrable in the sense that its Nijenhuis tensor vanishes (see (8.1) below).*

Condition (a) requires the real dimension of M to be even. Obviously, given the metric and one of the tensors J and Ω, we can immediately recover the other one by the formula (c).

Here are a few examples of Kähler manifolds which will be studied later on in these notes:

- $(\mathbb{C}^m, \langle\,,\,\rangle)$, where $\langle\,,\,\rangle$ denotes the canonical Hermitian metric $\langle\,,\,\rangle = \text{Re}(\sum dz_i d\bar{z}_i)$;
- oriented 2-dimensional Riemannian manifolds;
- the complex projective space $(\mathbb{C}P^m, FS)$ endowed with the Fubini–Study metric;
- projective manifolds, that is, submanifolds of $\mathbb{C}P^m$ defined by homogeneous polynomials in $\mathbb{C}^{m+1}$.

Kähler structures were introduced by Erich Kähler in his article [8] with the following motivation. Given any Hermitian metric h on a complex manifold, we can express the fundamental 2-form Ω in local holomorphic coordinates as follows:

$$\Omega = i \sum h_{\alpha\bar\beta} dz_\alpha \wedge d\bar z_\beta,$$

where

$$h_{\alpha\bar\beta} := h\left(\frac{\partial}{\partial z_\alpha}, \frac{\partial}{\partial \bar z_\beta}\right).$$

He then noticed that the condition $d\Omega = 0$ is equivalent (see Proposition 8.8 below) to the local existence of some function u such that

$$h_{\alpha\bar\beta} = \frac{\partial^2 u}{\partial z_\alpha \partial \bar z_\beta}.$$

In other words, the whole metric tensor is defined by a unique function! This remarkable (*bemerkenswert*) property of the metric allows one to obtain simple explicit expressions for the Ricci and curvature tensors, and "a long list of miracles occur then". The function u is called a *local Kähler potential*.

There is another remarkable property of Kähler metrics which, curiously, Kähler himself did not seem to have noticed. Recall that every point x in a Riemannian manifold has a local coordinate system x_i such that the metric osculates to the Euclidean metric to the order 2 at x. These special coordinate systems are the *normal coordinates* around each point. Now, on a complex manifold with Hermitian metric, the existence of normal *holomorphic* coordinates around each point is equivalent to the metric being Kähler (see Section 11.3 below).

Kähler manifolds have found many applications in various domains such as differential geometry, complex analysis, algebraic geometry or theoretical physics. To illustrate their importance let us make the following remark. With two exceptions (flat metrics and Joyce metrics in dimensions 7 and 8), the only known compact examples of irreducible Riemannian metrics satisfying Einstein's equations

$$R_{\alpha\beta} = 0$$

(Ricci-flat in modern language) are constructed on Kähler manifolds. Generic Ricci-flat Kähler manifolds, also called *Calabi–Yau manifolds*, will be studied later on in these notes.

7.2. Holomorphic functions

A function $F = f + ig : U \subset \mathbb{C} \to \mathbb{C}$ is called *holomorphic* if it satisfies the Cauchy–Riemann equations:

$$\frac{\partial f}{\partial x} = \frac{\partial g}{\partial y} \quad \text{and} \quad \frac{\partial f}{\partial y} + \frac{\partial g}{\partial x} = 0.$$

Let j denote the endomorphism of $\mathbb{R}^2$ corresponding to the multiplication by i on $\mathbb{C}$ via the identification of $\mathbb{R}^2$ with $\mathbb{C}$ given by $z = x + iy \mapsto (x, y)$. The endomorphism j can be expressed in the canonical basis as

$$j := \begin{pmatrix} 0 & -1 \\ 1 & 0 \end{pmatrix}.$$

If we view F as a real function $F : U \subset \mathbb{R}^2 \to \mathbb{R}^2$, its differential at some $p \in U$ is of course the linear map

$$(F_*)_p = \begin{pmatrix} \frac{\partial f}{\partial x}(p) & \frac{\partial f}{\partial y}(p) \\ \frac{\partial g}{\partial x}(p) & \frac{\partial g}{\partial y}(p) \end{pmatrix}.$$

Then it is easy to check that the Cauchy–Riemann relations are equivalent to the commutation relation $j \circ (F_*)_p = (F_*)_p \circ j, \ \forall \, p \in U$.

Similarly, we identify $\mathbb{C}^m$ with $\mathbb{R}^{2m}$ via

$$(z_1, \ldots, z_m) = (x_1 + iy_1, \ldots, x_m + iy_m) \mapsto (x_1, \ldots, x_m, y_1, \ldots, y_m),$$

and denote by j_m the endomorphism of $\mathbb{R}^{2m}$ corresponding to the multiplication by i on $\mathbb{C}^m$:

$$j_m := \begin{pmatrix} 0 & -I_m \\ I_m & 0 \end{pmatrix}. \tag{7.1}$$

A map $F : U \subset \mathbb{C}^n \to \mathbb{C}^m$ is *holomorphic* if and only if the differential F_* of F as real map $F : U \subset \mathbb{R}^{2n} \to \mathbb{R}^{2m}$ satisfies $j_m \circ (F_*)_p = (F_*)_p \circ j_n, \ \forall \, p \in U$.

7.3. Complex manifolds

A *complex manifold* of complex dimension m is a topological manifold $(M, \mathcal{U})$ whose atlas $(\phi_U)_{U \in \mathcal{U}}$ satisfies the following compatibility condition: for every intersecting $U, V \in \mathcal{U}$, the map between open sets of $\mathbb{C}^m$

$$\phi_{UV} := \phi_U \circ \phi_V^{-1}$$

is holomorphic. A pair (U, ϕ_U) is called a chart and the collection of all charts is called a *holomorphic structure*.

IMPORTANT EXAMPLE. The *complex projective space* $\mathbb{C}P^m$ can be defined as the set of complex lines of $\mathbb{C}^{m+1}$ (a line is a vector subspace of dimension one). If we define the equivalence relation $\sim$ on $\mathbb{C}^{m+1} \setminus \{0\}$ by

$$(z_0, \ldots, z_m) \sim (\alpha z_0, \ldots, \alpha z_m), \qquad \forall\, \alpha \in \mathbb{C}^*,$$

then $\mathbb{C}P^m = (\mathbb{C}^{m+1} \setminus \{0\})/\!\sim$. The equivalence class of $(z_0, \ldots, z_m)$ will be denoted by $[z_0 : \ldots : z_m]$. Consider the open cover U_i, $i = 0, \ldots, m$ of $\mathbb{C}P^m$ defined by

$$U_i := \{[z_0 : \ldots : z_m] \mid z_i \neq 0\}$$

and the maps $\phi_i : U_i \to \mathbb{C}^m$,

$$\phi_i([z_0 : \ldots : z_m]) = \left(\frac{z_0}{z_i}, \ldots, \frac{z_{i-1}}{z_i}, \frac{z_{i+1}}{z_i}, \ldots, \frac{z_m}{z_i} \right).$$

It is then an easy exercise to compute

$$\phi_i \circ \phi_j^{-1}(w_1, \ldots, w_m) = \left(\frac{w_1}{w_i}, \ldots, \frac{w_{i-1}}{w_i}, \frac{w_{i+1}}{w_i}, \ldots, \frac{w_j}{w_i}, \frac{1}{w_i}, \frac{w_{j+1}}{w_i}, \ldots, \frac{w_m}{w_i} \right),$$

which is obviously holomorphic on its domain of definition.

A function $F : M \to \mathbb{C}$ is called holomorphic if $F \circ \phi_U^{-1}$ is holomorphic for every $U \in \mathcal{U}$. This property is local. To check it in the neighbourhood of a point x it is enough to check it for a single $U \in \mathcal{U}$ containing x.

Since every holomorphic map between open sets of $\mathbb{C}^m$ is in particular a smooth map between open sets of $\mathbb{R}^{2m}$, every complex manifold M of complex dimension m defines a *real* $2m$-dimensional smooth manifold $M_{\mathbb{R}}$, which is the same as M as topological space. The converse does not hold, of course (not every smooth map is holomorphic), but it is nevertheless remarkable that the holomorphic structure of M is encoded in a single tensor of the real manifold $M_{\mathbb{R}}$, which is a field of endomorphisms of the tangent bundle defined as follows. For every $X \in T_x M_{\mathbb{R}}$, choose $U \in \mathcal{U}$ containing x and define

$$J_U(X) = (\phi_U)_*^{-1} \circ j_m \circ (\phi_U)_*(X).$$

If we take some other $V \in \mathcal{U}$ containing x, then $\phi_{VU} = \phi_V \circ \phi_U^{-1}$ is holomorphic, and $\phi_V = \phi_{VU} \circ \phi_U$, so

$$\begin{aligned}
J_V(X) &= (\phi_V)_*^{-1} \circ j_m \circ (\phi_V)_*(X) = (\phi_V)_*^{-1} \circ j_m \circ (\phi_{VU})_* \circ (\phi_U)_*(X) \\
&= (\phi_V)_*^{-1} \circ (\phi_{VU})_* \circ j_m \circ (\phi_U)_*(X) = (\phi_U)_*^{-1} \circ j_m \circ (\phi_U)_*(X) \\
&= J_U(X),
\end{aligned}$$

showing that J_U does not depend on U. Their collection is thus a well-defined tensor J on $M_{\mathbb{R}}$ satisfying $J^2 = -\mathrm{Id}$.

DEFINITION 7.2. *A* $(1,1)$-*tensor J on a smooth (real) manifold M which satisfies $J^2 = -\mathrm{Id}$ is called an* almost complex structure. *The pair (M, J) is then referred to as an* almost complex manifold.

A complex manifold is thus in a canonical way an almost complex manifold. The converse is not true in general, but it holds under some integrability condition (see Theorem 7.4 below). From now on we will always identify a complex manifold M with its underlying real manifold $M_{\mathbb{R}}$, equipped with the tensor J.

7.4. The complexified tangent bundle

Let (M, J) be an almost complex manifold. We would like to diagonalize the endomorphism J. In order to do so, we have to complexify the tangent space. Define

$$TM^{\mathbb{C}} := TM \otimes_{\mathbb{R}} \mathbb{C}.$$

We extend all real endomorphisms and differential operators from TM to $TM^{\mathbb{C}}$ by $\mathbb{C}$-linearity. Let $T^{1,0}M$ (resp. $T^{0,1}M$) denote the eigenbundle of $TM^{\mathbb{C}}$ corresponding to the eigenvalue i (resp. $-i$) of J. The following algebraic lemma is an easy exercise.

LEMMA 7.3. *One has*

$$T^{1,0}M = \{X - iJX \mid X \in TM\}, \qquad T^{0,1}M = \{X + iJX \mid X \in TM\},$$

and $TM^{\mathbb{C}} = T^{1,0}M \oplus T^{0,1}M$.

The famous Newlander–Nirenberg theorem can be stated as follows:

THEOREM 7.4. *Let (M, J) be an almost complex manifold. The almost complex structure J comes from a holomorphic structure if and only if the distribution $T^{0,1}M$ is integrable.*

PROOF. We will only prove here the "only if" part. The interested reader can find the proof of the hard part for example in [6].

Suppose that J comes from a holomorphic structure on M. Consider a holomorphic chart (U, ϕ_U) and let $z_\alpha = x_\alpha + iy_\alpha$ be the αth component of ϕ_U. If $\{e_1, \ldots, e_{2m}\}$ denotes the standard basis of $\mathbb{R}^{2m}$, we have by definition:

$$\frac{\partial}{\partial x_\alpha} = (\phi_U)_*^{-1}(e_\alpha) \qquad \text{and} \qquad \frac{\partial}{\partial y_\alpha} = (\phi_U)_*^{-1}(e_{m+\alpha}).$$

Moreover, $j_m(e_\alpha) = e_{m+\alpha}$, so we obtain directly from the definition

$$J\left(\frac{\partial}{\partial x_\alpha}\right) = \frac{\partial}{\partial y_\alpha}. \tag{7.2}$$

We now make the following notations

$$\frac{\partial}{\partial z_\alpha} := \frac{1}{2}\left(\frac{\partial}{\partial x_\alpha} - i\frac{\partial}{\partial y_\alpha}\right), \qquad \frac{\partial}{\partial \bar{z}_\alpha} := \frac{1}{2}\left(\frac{\partial}{\partial x_\alpha} + i\frac{\partial}{\partial y_\alpha}\right).$$

From Lemma 7.3 and (7.2) we obtain immediately that $\partial/\partial z_\alpha$ and $\partial/\partial \bar{z}_\alpha$ are local sections of $T^{1,0}M$ and $T^{0,1}M$ respectively. They form moreover a local basis at each point of U. Let now Z and W be two local sections of

$T^{0,1}M$ written as $Z = \sum Z_\alpha(\partial/\partial\bar{z}_\alpha)$, $W = \sum W_\alpha(\partial/\partial\bar{z}_\alpha)$ in this local basis. A direct calculation then gives

$$[Z, W] = \sum_{\alpha,\beta=1}^{m} Z_\alpha \frac{\partial W_\beta}{\partial\bar{z}_\alpha} \frac{\partial}{\partial\bar{z}_\beta} - \sum_{\alpha,\beta=1}^{m} W_\alpha \frac{\partial Z_\beta}{\partial\bar{z}_\alpha} \frac{\partial}{\partial\bar{z}_\beta},$$

which is clearly a local section of $T^{0,1}M$. $\qquad\qquad\square$

An almost complex structure arising from a holomorphic structure is called a *complex structure*.

REMARK. The existence of local coordinates satisfying (7.2) is actually the key point of the hard part of the theorem. Once we have such coordinates, it is easy to show that the transition functions are holomorphic: suppose that (u_α, v_α) is another local system of coordinates, satisfying

$$\frac{\partial}{\partial v_\alpha} = J\frac{\partial}{\partial u_\alpha}.$$

We then have

$$\frac{\partial}{\partial x_\alpha} = \sum_{\beta=1}^{m} \frac{\partial u_\beta}{\partial x_\alpha} \frac{\partial}{\partial u_\beta} + \sum_{\beta=1}^{m} \frac{\partial v_\beta}{\partial x_\alpha} \frac{\partial}{\partial v_\beta} \qquad (7.3)$$

and

$$\frac{\partial}{\partial y_\alpha} = \sum_{\beta=1}^{m} \frac{\partial u_\beta}{\partial y_\alpha} \frac{\partial}{\partial u_\beta} + \sum_{\beta=1}^{m} \frac{\partial v_\beta}{\partial y_\alpha} \frac{\partial}{\partial v_\beta}. \qquad (7.4)$$

Applying J to (7.3) and comparing to (7.4) yields

$$\frac{\partial u_\beta}{\partial x_\alpha} = \frac{\partial v_\beta}{\partial y_\alpha} \qquad \text{and} \qquad \frac{\partial u_\beta}{\partial y_\alpha} = -\frac{\partial v_\beta}{\partial x_\alpha},$$

thus showing that the transition functions are holomorphic.

7.5. Exercises

(1) Prove Lemma 7.3.

(2) Let $A + iB \in \mathrm{Gl}_m(\mathbb{C})$. Compute the product

$$\begin{pmatrix} I_m & 0 \\ -iI_m & I_m \end{pmatrix} \begin{pmatrix} A & B \\ -B & A \end{pmatrix} \begin{pmatrix} I_m & 0 \\ iI_m & I_m \end{pmatrix}$$

and use this computation to prove that for every invertible matrix $A + iB \in \mathrm{Gl}_m(\mathbb{C})$, the determinant of the real $2m \times 2m$ matrix

$$\begin{pmatrix} A & B \\ -B & A \end{pmatrix}$$

is strictly positive.

(3) Show that every almost complex manifold is orientable.

(4) Let $\alpha > 1$ be some real number. Let Γ be the subgroup of $\mathrm{Gl}_m(\mathbb{C})$ generated by αI_m. Show that Γ acts freely and properly discontinuously on $\mathbb{C}^m \setminus \{0\}$ (see [10], p. 43). Use this to prove that $S^1 \times S^{2m-1}$ carries a complex structure.

(5) Show that every holomorphic function defined on a compact complex manifold is constant. *Hint:* Use the maximum principle.

(6) (*The complex Grassmannian*) Let $\mathrm{Gr}_p(\mathbb{C}^m)$ denote the set of all p-dimensional vector subspaces of $\mathbb{C}^m$. Show that $\mathrm{Gr}_p(\mathbb{C}^m)$ has the structure of a complex manifold of dimension $p(m-p)$. *Hint:* If $\{e_i\}$ denotes the canonical basis of $\mathbb{C}^m$, and $I := \{i_1, \ldots, i_{m-p}\}$ is a subset of $\{1, \ldots, m\}$, let U_I be the subset of $\mathrm{Gr}_p(\mathbb{C}^m)$ consisting of all p-dimensional vector subspaces of $\mathbb{C}^m$ supplementary to the vector subspace spanned by $\{e_{i_1}, \ldots, e_{i_{m-p}}\}$. Show that U_I can be naturally identified with the set of $p \times (m-p)$ matrices, and that their collection defines a holomorphic atlas on $\mathrm{Gr}_p(\mathbb{C}^m)$.

CHAPTER 8

Holomorphic forms and vector fields

8.1. Decomposition of the (complexified) exterior bundle

Let (M, J) be an almost complex manifold. We now turn our attention to exterior forms and introduce the complexified exterior bundle $\Lambda_{\mathbb{C}}^* M = \Lambda^* M \otimes_{\mathbb{R}} \mathbb{C}$. The sections of $\Lambda_{\mathbb{C}}^* M$ can be viewed as complex-valued forms or as formal sums $\omega + i\tau$, where ω and τ are usual real forms on M.

We define the following two subbundles of $\Lambda_{\mathbb{C}}^1 M$:

$$\Lambda^{1,0} M := \{\xi \in \Lambda_{\mathbb{C}}^1 M \mid \xi(Z) = 0, \ \forall \, Z \in T^{0,1} M\}$$

and

$$\Lambda^{0,1} M := \{\xi \in \Lambda_{\mathbb{C}}^1 M \mid \xi(Z) = 0, \ \forall \, Z \in T^{1,0} M\}.$$

The sections of these subbundles are called forms of type (1,0) or forms of type (0,1) respectively. Lemma 7.3 yields the following:

LEMMA 8.1. *One has*

$$\Lambda^{1,0} M = \{\omega - i\omega \circ J \mid \omega \in \Lambda^1 M\}, \qquad \Lambda^{0,1} M = \{\omega + i\omega \circ J \mid \omega \in \Lambda^1 M\}$$

and $\Lambda_{\mathbb{C}}^1 M = \Lambda^{1,0} M \oplus \Lambda^{0,1} M$.

Let us denote the kth exterior power of $\Lambda^{1,0}$ (resp. $\Lambda^{0,1}$) by $\Lambda^{k,0}$ (resp. $\Lambda^{0,k}$) and let $\Lambda^{p,q}$ denote the tensor product $\Lambda^{p,0} \otimes \Lambda^{0,q}$. The exterior power of a direct sum of vector spaces can be described as follows

$$\Lambda^k(E \oplus F) \simeq \bigoplus_{i=0}^{k} \Lambda^i E \otimes \Lambda^{k-i} F.$$

From Lemma 8.1 we then get

$$\Lambda_{\mathbb{C}}^k M \simeq \bigoplus_{p+q=k} \Lambda^{p,q} M.$$

Sections of $\Lambda^{p,q} M$ are called forms of type (p, q) and the space of forms of type (p, q) is denoted by $\Omega^{p,q} M$. It is easy to check that a complex-valued k-form ω belongs to $\Omega^{k,0} M$ if and only if $Z \lrcorner \omega = 0$ for all $Z \in T^{0,1} M$. More generally, a complex k-form belongs to $\Lambda^{p,q} M$ if and only if it vanishes whenever applied to $p + 1$ vectors from $T^{1,0} M$ or to $q + 1$ vectors from $T^{0,1} M$.

If J is a complex structure, we can describe these spaces in terms of a local holomorphic coordinate system. Let $z_\alpha = x_\alpha + iy_\alpha$ be the αth coordinate of some ϕ_U. Extending the exterior derivative on complex-valued functions by $\mathbb{C}$-linearity defines complex-valued 1-forms $dz_\alpha = dx_\alpha + idy_\alpha$ and $d\bar{z}_\alpha = dx_\alpha - idy_\alpha$. Then $\{dz_1, \ldots, dz_m\}$ and $\{d\bar{z}_1, \ldots, d\bar{z}_m\}$ are local bases for $\Lambda^{1,0}M$ and $\Lambda^{0,1}M$ respectively, and a local basis for $\Lambda^{p,q}M$ is given by

$$\{dz_{i_1} \wedge \cdots \wedge dz_{i_p} \wedge d\bar{z}_{j_1} \wedge \cdots \wedge d\bar{z}_{j_q}, \; i_1 < \cdots < i_p, \; j_1 < \cdots < j_q\}.$$

To every almost complex structure J one can associate a $(2,1)$-tensor N^J called the Nijenhuis tensor, defined by

$$N^J(X,Y) = [X,Y] + J[JX,Y] + J[X,JY] - [JX,JY], \qquad (8.1)$$

for all $X,Y \in \mathcal{X}(M)$.

PROPOSITION 8.2. *Let J be an almost complex structure on a real $2m$-dimensional manifold M. The following statements are equivalent:*

(1) *J is a complex structure.*
(2) *$T^{0,1}M$ is formally integrable, that is, $[Z,W] \in \Gamma(T^{0,1}M)$, $\forall \, Z,W \in \Gamma(T^{0,1}M)$.*
(3) *$d(\Omega^{1,0}M) \subset \Omega^{2,0}M \oplus \Omega^{1,1}M$.*
(4) *$d(\Omega^{p,q}M) \subset \Omega^{p+1,q}M \oplus \Omega^{p,q+1}M \; \forall \, 0 \leq p,q \leq m$.*
(5) *$N^J = 0$.*

PROOF. $(1) \Longleftrightarrow (2)$ is given by Theorem 7.4.

$(2) \Longleftrightarrow (3)$. Let ω be a section of $\Lambda^{1,0}M$. The $\Lambda^{0,2}M$-component of $d\omega$ vanishes if and only if $d\omega(Z,W) = 0$ for all $Z,W \in T^{0,1}M$. Extend Z and W to local sections of $T^{0,1}M$ and write

$$d\omega(Z,W) = \partial_Z(\omega(W)) - \partial_W(\omega(Z)) - \omega([Z,W]) = -\omega([Z,W]).$$

Thus

$$d\omega(Z,W) = 0, \quad \forall \, Z,W \in T^{0,1}M, \; \forall \, \omega \in \Omega^{1,0}M$$
$$\Longleftrightarrow \quad \omega([Z,W]) = 0, \quad \forall \, Z,W \in \Gamma(T^{0,1}M), \; \forall \, \omega \in \Omega^{1,0}M$$
$$\Longleftrightarrow \quad [Z,W] \in \Gamma(T^{0,1}M), \quad \forall \, Z,W \in \Gamma(T^{0,1}M).$$

$(3) \Longleftrightarrow (4)$. Suppose that (3) holds. By complex conjugation we get immediately $d(\Omega^{0,1}M) \subset \Omega^{0,2}M \oplus \Omega^{1,1}M$. It is then enough to apply the Leibniz rule to any section of $\Lambda^{p,q}M$, locally written as a sum of decomposable elements $\omega_1 \wedge \cdots \wedge \omega_p \wedge \bar{\tau}_1 \wedge \cdots \wedge \bar{\tau}_q$, where $\omega_i \in \Omega^{1,0}M$ and $\bar{\tau}_i \in \Omega^{0,1}M$. The reverse implication is obvious.

$(2) \Longleftrightarrow (5)$. Let $X,Y \in \mathcal{X}(M)$ be local vector fields and let Z denote the bracket $Z := [X + iJX, Y + iJY]$. An easy calculation gives $Z - iJZ = N^J(X,Y) - iJN^J(X,Y)$. Thus $Z \in T^{0,1}M \Longleftrightarrow N^J(X,Y) = 0$, which proves that $T^{0,1}M$ is integrable if and only if $N^J \equiv 0$. $\qquad \square$

8.2. Holomorphic objects on complex manifolds

In this section (M, J) will denote a complex manifold of complex dimension m. We start with the following characterization of holomorphic functions.

LEMMA 8.3. *Let $f : M \to \mathbb{C}$ be a smooth complex-valued function on M. The following assertions are equivalent:*

(1) *f is holomorphic.*
(2) *$\partial_Z f = 0, \ \forall \, Z \in T^{0,1} M$.*
(3) *df is a form of type $(1, 0)$.*

PROOF. $(2) \Longleftrightarrow (3)$. $df \in \Omega^{1,0} M \Longleftrightarrow df(Z) = 0, \ \forall \, Z \in T^{0,1} M \Longleftrightarrow \partial_Z f = 0, \ \forall \, Z \in T^{0,1} M$.

$(1) \Longleftrightarrow (3)$. The function f is holomorphic if and only if $f \circ \phi_U^{-1}$ is holomorphic for every holomorphic chart (U, ϕ_U), which is equivalent to $f_* \circ (\phi_U)_*^{-1} \circ j_m = i f_* \circ (\phi_U)_*^{-1}$, that is, $f_* \circ J = i f_*$. This last equation just means that for every real vector X, $df(JX) = i df(X)$, that is, $i df(X + iJX) = 0, \ \forall \, X \in TM$, which is equivalent to $df \in \Omega^{1,0} M$. □

Using Proposition 8.2, for every fixed (p, q) we define the differential operators $\partial : \Omega^{p,q} M \to \Omega^{p+1,q} M$ and $\bar{\partial} : \Omega^{p,q} M \to \Omega^{p,q+1} M$ by $d = \partial + \bar{\partial}$.

LEMMA 8.4. *The following identities hold:*
$$\partial^2 = 0, \qquad \bar{\partial}^2 = 0, \qquad \partial\bar{\partial} + \bar{\partial}\partial = 0.$$

PROOF. We have $0 = d^2 = (\partial + \bar{\partial})^2 = \partial^2 + \bar{\partial}^2 + (\partial\bar{\partial} + \bar{\partial}\partial)$, and the three operators in the last term take values in different subbundles of $\Lambda_{\mathbb{C}}^* M$. □

DEFINITION 8.5. *A vector field Z in $\Gamma(T^{1,0} M)$ is called* holomorphic *if $\partial_Z f$ is holomorphic for every locally defined holomorphic function f. A p-form ω of type $(p, 0)$ is called* holomorphic *if $\bar{\partial}\omega = 0$.*

DEFINITION 8.6. *A real vector field X is called* real holomorphic *if its $(1, 0)$ component $X - iJX$ is a holomorphic vector field.*

LEMMA 8.7. *Let X be a real vector field on a complex manifold (M, J). The following assertions are equivalent:*

(1) *X is real holomorphic.*
(2) *$\mathcal{L}_X J = 0$.*
(3) *The flow of X consists of holomorphic transformations of M.*

Although not explicitly stated, the reader might have guessed that a smooth map $f : (M, J_1) \to (N, J_2)$ between two complex manifolds is called holomorphic if its differential commutes with the complex structures at each point: $f_* \circ J_1 = J_2 \circ f_*$.

PROOF. The equivalence of the last two assertions is tautological. In order to prove the equivalence of the first two assertions, we first notice that a complex vector field Z is of type $(0,1)$ if and only if $\partial_Z f = 0$ for every locally defined holomorphic function f. Suppose that X is real holomorphic and let Y be an arbitrary vector field and f a local holomorphic function. As $\partial_{(X+iJX)} f = 0$, we have $\partial_{(X-iJX)} f = 2\partial_X f$. By definition, $\partial_X f$ is then holomorphic, so by Lemma 8.3 we get $\partial_{(Y+iJY)}(\partial_X f) = 0$ and $\partial_{(Y+iJY)} f = 0$. In particular, $\partial_{[Y+iJY,X]}(f) = 0$. As this holds for every holomorphic f, $[Y + iJY, X]$ has to be of type $(0,1)$, that is, $[JY, X] = J[Y, X]$. Hence $(\mathcal{L}_X J)(Y) = \mathcal{L}_X(JY) - J(\mathcal{L}_X Y) = [X, JY] - J[X, Y] = 0$ for all vector fields Y, i.e. $\mathcal{L}_X J = 0$. The converse is similar and left to the reader.   □

REMARK. If M is a compact complex manifold, its automorphism group (consisting of holomorphic diffeomorphisms) is a Lie group, whose Lie algebra is just the space of real holomorphic vector fields.

We close this section with the following important result:

PROPOSITION 8.8. (The local $i\partial\bar{\partial}$-lemma) *Let $\omega \in \Lambda^{1,1} M \cap \Lambda^2 M$ be a real 2-form of type $(1,1)$ on a complex manifold M. Then ω is closed if and only if every point $x \in M$ has an open neighbourhood U such that the restriction of ω to U equals $i\partial\bar{\partial}u$ for some real function u on U.*

PROOF. One implication is clear from Lemma 8.4:

$$d(i\partial\bar{\partial}) = i(\partial + \bar{\partial})\partial\bar{\partial} = i(\partial^2\bar{\partial} - \partial\bar{\partial}^2) = 0.$$

The other implication is more delicate and needs the following counterpart of the Poincaré Lemma (see [3], p. 25 for a proof):

LEMMA 8.9. (Dolbeault Lemma) *A $\bar{\partial}$-closed $(0,1)$-form is locally $\bar{\partial}$-exact.*

Let ω be a closed real form of type $(1,1)$. From the Poincaré Lemma, there exists locally a real 1-form τ with $d\tau = \omega$. Let $\tau = \tau^{1,0} + \tau^{0,1}$ be the decomposition of τ in forms of type (1,0) and (0,1). Clearly, $\tau^{1,0} = \overline{\tau^{0,1}}$. Comparing types in the equality

$$\omega = d\tau = \bar{\partial}\tau^{0,1} + (\partial\tau^{0,1} + \bar{\partial}\tau^{1,0}) + \partial\tau^{1,0},$$

we get $\bar{\partial}\tau^{0,1} = 0$ and $\omega = (\partial\tau^{0,1} + \bar{\partial}\tau^{1,0})$. By the Dolbeault Lemma, there exists a local function f such that $\tau^{0,1} = \bar{\partial}f$. By complex conjugation we get $\tau^{1,0} = \partial\bar{f}$, whence $\omega = (\partial\tau^{0,1} + \bar{\partial}\tau^{1,0}) = \partial\bar{\partial}f + \bar{\partial}\partial\bar{f} = i\partial\bar{\partial}(2\mathrm{Im}(f))$, and the proposition follows, with $u := 2\mathrm{Im}(f)$.   □

8.3. Exercises

(1) Prove Lemma 8.1.

(2) Prove that the object defined by formula (8.1) is indeed a tensor.

(3) Show that an almost complex structure on a real 2-dimensional manifold is always integrable.

(4) Show that $\{dz_\alpha\}$ and $\{\partial/\partial z_\alpha\}$ are dual bases of $\Lambda^{1,0}M$ and $T^{1,0}M$ at each point of the local coordinate system.

(5) Show that a 2-form ω is of type (1,1) if and only if $\omega(X,Y) = \omega(JX, JY)$, $\forall\, X, Y \in TM$.

(6) Let M be a complex manifold with local holomorphic coordinates $\{z_\alpha\}$.
   - Prove that a local vector field of type (1,0), $Z = \sum Z_\alpha(\partial/\partial z_\alpha)$ is holomorphic if and only if Z_α are holomorphic functions.
   - Prove that a local form of type (1,0), $\omega = \sum \omega_\alpha dz_\alpha$ is holomorphic if and only if ω_α are holomorphic functions.

(7) If X is a real holomorphic vector field on a complex manifold, prove that JX has the same property.

(8) Prove the converse in Lemma 8.7.

(9) Show that in every local coordinate system one has

$$\partial f = \sum_{\alpha=1}^{m} \frac{\partial f}{\partial z_\alpha} dz_\alpha \quad \text{and} \quad \bar{\partial} f = \sum_{\alpha=1}^{m} \frac{\partial f}{\partial \bar{z}_\alpha} d\bar{z}_\alpha.$$

(10) Let M be a real manifold, and let T be a complex subbundle of $\Lambda^1_{\mathbb{C}}M$ such that $T \oplus \Lambda^1 M = \Lambda^1_{\mathbb{C}}M$. Show that there exists a unique almost complex structure J on M such that $T = \Lambda^{1,0}M$ with respect to J.

CHAPTER 9

Complex and holomorphic vector bundles

9.1. Holomorphic vector bundles

Let M be a complex manifold and let $\pi : E \to M$ be a complex vector bundle over M (i.e. each fibre $\pi^{-1}(x)$ is a complex vector space). E is called a *holomorphic vector bundle* if there exists a trivialization with holomorphic transition functions. More precisely, there exists an open cover $\mathcal{U}$ of M and for each $U \in \mathcal{U}$ a diffeomorphism $\psi_U : \pi^{-1}(U) \to U \times \mathbb{C}^k$ such that

- the following diagram commutes:

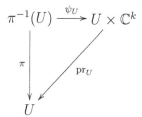

- for every intersecting U and V one has $\psi_U \circ \psi_V^{-1}(x, v) = (x, g_{UV}(x)v)$, where $g_{UV} : U \cap V \to \mathrm{Gl}_k(\mathbb{C}) \subset \mathbb{C}^{k^2}$ are holomorphic functions.

EXAMPLES. 1. The tangent bundle of a complex manifold M^{2m} is holomorphic. To see this, take a holomorphic atlas (U, ϕ_U) on M and define $\psi_U : TM|_U \to U \times \mathbb{C}^m$ by $\psi_U(X_x) = (x, (\phi_U)_*(X))$. The transition functions $g_{UV} = (\phi_U)_* \circ (\phi_V)_*^{-1}$ are then clearly holomorphic.

2. The cotangent bundle, and more generally the bundles $\Lambda^{p,0} M$ are holomorphic. Indeed, using again a holomorphic atlas of the manifold one can trivialize locally $\Lambda^{p,0} M$ and the chain rule

$$dz_{\alpha_1} \wedge \cdots \wedge dz_{\alpha_p} = \sum_{\beta_1, .., \beta_p} \frac{\partial z_{\alpha_1}}{\partial w_{\beta_1}} \cdots \frac{\partial z_{\alpha_p}}{\partial w_{\beta_p}} dw_{\beta_1} \wedge \cdots \wedge dw_{\beta_p}$$

shows that the transition functions are holomorphic.

For every holomorphic bundle E one defines the bundles $\Lambda^{p,q}(E) := \Lambda^{p,q} M \otimes E$ of E-valued forms on M of type (p, q). The space of sections of $\Lambda^{p,q} E$ is denoted by $\Omega^{p,q}(E)$. We define the $\bar{\partial}$-operator $\bar{\partial} : \Omega^{p,q}(E) \to \Omega^{p,q+1}(E)$ in the following way. If a section σ of $\Lambda^{p,q}(E)$ is given by $\sigma = (\omega_1, \ldots, \omega_k)$ in some local trivialization (where ω_i are local (p, q)-forms), we define $\bar{\partial}\sigma := (\bar{\partial}\omega_1, \ldots, \bar{\partial}\omega_k)$. Suppose that σ is written $\sigma = (\tau_1, \ldots, \tau_k)$ in some other trivialization of E. Then one has $\tau_j = \sum_{l=1}^{k} g_{jl}\omega_l$ for some holomorphic functions g_{jl}, thus $\bar{\partial}\tau_j = \sum_{l=1}^{k} g_{jl}\bar{\partial}\omega_l$, showing that $\bar{\partial}\sigma$ does not depend on the chosen trivialization. By construction one has $\bar{\partial}^2 = 0$ and $\bar{\partial}$ satisfies the Leibniz rule:

$$\bar{\partial}(\omega \wedge \sigma) = (\bar{\partial}\omega) \wedge \sigma + (-1)^{p+q}\omega \wedge (\bar{\partial}\sigma), \qquad \forall\, \omega \in \Omega^{p,q}M,\ \sigma \in \Omega^{r,s}(E).$$

Notice that the bundles $\Lambda^{p,q}M$ *are not* holomorphic bundles for $q \neq 0$.

9.2. Holomorphic structures

A *pseudo-holomorphic structure* on a complex vector bundle E is an operator $\bar{\partial} : \Omega^{p,q}(E) \to \Omega^{p,q+1}(E)$ satisfying the Leibniz rule. If, moreover, $\bar{\partial}^2 = 0$, then $\bar{\partial}$ is called a *holomorphic structure*.

A section σ in a pseudo-holomorphic vector bundle $(E, \bar{\partial})$ is called *holomorphic* if $\bar{\partial}\sigma = 0$.

LEMMA 9.1. *A pseudo-holomorphic vector bundle $(E, \bar{\partial})$ of rank k is holomorphic if and only if each $x \in M$ has an open neighbourhood U and k holomorphic sections σ_i of E over U such that $\{\sigma_i(x)\}$ form a basis of E_x.*

PROOF. If E is holomorphic, one can define for every local holomorphic trivialization (U, ψ_U) a local basis of holomorphic sections by $\sigma_i(x) := \psi_U^{-1}(x, e_i)$, $\forall\, x \in U$. Conversely, every local basis of holomorphic sections defines a local trivialization of E. If $\{\sigma_i\}$ and $\{\tilde{\sigma}_i\}$ are two such holomorphic bases, corresponding to two local trivializations (U, ψ_U) and (V, ψ_V), we can write $\sigma_i = \sum g_{ij}\tilde{\sigma}_j$ for some smooth functions g_{ij} defined on $U \cap V$. Applying $\bar{\partial}$ and using the Leibniz rule in this expression yield $\bar{\partial}g_{ij} = 0$, hence the transition functions $g_{UV} = (g_{ij})$ are holomorphic. $\square$

THEOREM 9.2. *A complex vector bundle E is holomorphic if and only if it has a holomorphic structure $\bar{\partial}$.*

PROOF. The "only if" part follows directly from the discussion above. Suppose, conversely, that E is a complex bundle over M of rank k with holomorphic structure $\bar{\partial}$ satisfying the Leibniz rule and $\bar{\partial}^2 = 0$. In order to show that E is holomorphic, it is enough to show, using Lemma 9.1, that one can trivialize E around each $x \in M$ by holomorphic sections. Let $\{\sigma_1, \ldots, \sigma_k\}$ be local sections of E which form a basis of E over some open set U containing x. We define local $(0,1)$-forms τ_{ij} on U by the formula

$$\bar{\partial}\sigma_i = \sum_{j=1}^{k} \tau_{ij} \otimes \sigma_j.$$

The condition $\bar\partial^2 = 0$, together with the Leibniz rule, yield

$$0 = \bar\partial^2 \sigma_i = \sum_{j=1}^{k} \bar\partial\tau_{ij} \otimes \sigma_j - \sum_{j,l=1}^{k} \tau_{il} \wedge \tau_{lj} \otimes \sigma_j,$$

whence

$$\bar\partial\tau_{ij} = \sum_{l=1}^{k} \tau_{il} \wedge \tau_{lj}, \quad \forall\, 1 \le i,j \le k. \tag{9.1}$$

From now on we will use the summation convention on repeating subscripts. Suppose one can find a map $f : U' \to \mathrm{Gl}_k(\mathbb{C})$, $f = (f_{ij})$ such that

$$0 = \bar\partial f_{ij} + f_{il}\tau_{lj}, \quad \forall\, 1 \le i,j \le k, \tag{9.2}$$

for some open subset U' of U containing x. It is then easy to check that the local sections s_j of E over U' defined by $s_j := f_{jl}\sigma_l$ are holomorphic:

$$\bar\partial s_j = \bar\partial f_{jl} \otimes \sigma_l + f_{jr}\tau_{rl} \otimes \sigma_l = 0.$$

The theorem thus follows from the next lemma. □

LEMMA 9.3. *Suppose that $\tau := (\tau_{ij})$ is a $\mathfrak{gl}_k(\mathbb{C})$-valued $(0,1)$-form on U satisfying $\bar\partial\tau = \tau \wedge \tau$, or equivalently (9.1). Then for every $x \in U$ there exists some open subset U' of U containing x and a map $f : U' \to \mathrm{Gl}_k(\mathbb{C})$, $f = (f_{ij})$ such that $\bar\partial f + f\tau = 0$, in other words, such that (9.2) holds.*

PROOF. The main idea is to define an almost complex structure locally on $U \times \mathbb{C}^k$ using τ, to show that its integrability is equivalent to (9.1), and finally to obtain f as the matrix of some frame defined by τ in terms of holomorphic coordinates given by the theorem of Newlander–Nirenberg.

We denote by N the product $U \times \mathbb{C}^k$. We may suppose that U is an open subset of $\mathbb{C}^m$ with holomorphic coordinates z_α and denote by w_i the coordinates in $\mathbb{C}^k$.

It is easy to check that any complement T of $\Lambda^1 N$ in the complexified bundle $\Lambda^1_{\mathbb{C}} N$, with $iT = T$, defines an almost complex structure on N, such that T becomes the space of (1,0)-forms on N (see Exercise (10) of Section 8.3).

Consider the subbundle T of $\Lambda^1 N \otimes \mathbb{C}$ generated by the 1-forms

$$\{dz_\alpha,\ dw_i - \tau_{il}w_l \mid 1 \le \alpha \le m,\ 1 \le i \le k\}.$$

We claim that the almost complex structure induced on N by T is integrable. By Proposition 8.2, we have to show that $d\Gamma(T) \subset \Gamma(T \wedge \Lambda^1_{\mathbb{C}} N)$. It is enough to check this on the local basis defining T. Clearly $d(dz_\alpha) = 0$ and from (9.1) we get

$$
\begin{aligned}
d(dw_i - \tau_{il} w_l) &= -\partial \tau_{il} w_l - \bar{\partial} \tau_{il} w_l + \tau_{il} \wedge dw_l \\
&= -\partial \tau_{il} w_l - \tau_{is} \wedge \tau_{sl} w_l + \tau_{is} \wedge dw_s \\
&= -\partial \tau_{il} w_l + \tau_{is} \wedge (dw_s - \tau_{sl} w_l),
\end{aligned}
$$

which clearly is a section of $\Gamma(T \wedge \Lambda_{\mathbb{C}}^1 N)$. We now use the Newlander–Nirenberg theorem and complete the family $\{z_\alpha\}$ to a local holomorphic coordinate system $\{z_\alpha, u_l\}$ on some smaller neighbourhood U' of x. Since du_l are sections of T, we can find functions F_{li} and $F_{l\alpha}$, $1 \le i, l \le k$, $1 \le \alpha \le m$ such that

$$
du_l = F_{li}(dw_i - \tau_{ik} w_k) + F_{l\alpha} dz_\alpha.
$$

We apply the exterior derivative to this system and get

$$
0 = dF_{li} \wedge (dw_i - \tau_{ik} w_k) + F_{li}(-d\tau_{ik} w_k + \tau_{ik} \wedge dw_k) + dF_{l\alpha} \wedge dz_\alpha.
$$

We evaluate this last equality for $w_i = 0$, and get

$$
0 = dF_{lk}(z, 0) \wedge dw_k + F_{li}(z, 0)\tau_{ik} \wedge dw_k + dF_{l\alpha} \wedge dz_\alpha. \tag{9.3}
$$

If we denote $f_{lk}(z) := F_{lk}(z, 0)$, then the $\Lambda^{0,1} U'$-part of $dF_{lk}(z, 0)$ is just $\bar{\partial} f_{lk}$. Therefore, the vanishing of the $\Lambda^{0,1} U' \otimes \Lambda^{1,0} \mathbb{C}^k$-components of (9.3) just reads

$$
0 = \bar{\partial} f_{lk} + f_{li} \tau_{ik}.
$$

□

The above proof was taken from [12].

9.3. The canonical bundle of $\mathbb{C}P^m$

Let (M, J) be a complex manifold of (complex) dimension m. The complex line bundle $K_M := \Lambda^{m,0} M$ is called the *canonical bundle* of M. We already noticed that K_M has a holomorphic structure.

On the complex projective space there is some distinguished holomorphic line bundle called the *tautological line bundle*. It is defined as the complex line bundle $\pi : L \to \mathbb{C}P^m$ whose fibre $L_{[z]}$ over some point $[z] \in \mathbb{C}P^m$ is the complex line $\langle z \rangle$ in $\mathbb{C}^{m+1}$.

We consider the canonical holomorphic charts (U_α, ϕ_α) on $\mathbb{C}P^m$ and the local trivializations $\psi_\alpha : \pi^{-1}(U_\alpha) \to U_\alpha \times \mathbb{C}$ of L defined by $\psi_\alpha([z], w) = ([z], w_\alpha)$. It is an easy exercise to compute the transition functions:

$$
\psi_\alpha \circ \psi_\beta^{-1}([z], \lambda) = ([z], g_{\alpha\beta}([z])\lambda), \quad \text{where } g_{\alpha\beta}([z]) = \frac{z_\alpha}{z_\beta}.
$$

This shows that the tautological bundle of $\mathbb{C}P^m$ is holomorphic.

PROPOSITION 9.4. *The canonical bundle of $\mathbb{C}P^m$ is isomorphic to the $(m+1)$th power of the tautological bundle.*

PROOF. A trivialization for $p : \Lambda^{m,0}\mathbb{C}P^m \to \mathbb{C}P^m$ is given by $(\phi_\alpha^*)^{-1} :$ $p^{-1}(U_\alpha) \to U_\alpha \times \Lambda^{m,0}\mathbb{C}^m$, so the transition functions are $h_{\alpha\beta} := (\phi_\alpha^*)^{-1} \circ (\phi_\beta^*)$. Let now $\omega := dw_1 \wedge \cdots \wedge dw_m$ be the canonical generator of $\Lambda^{m,0}\mathbb{C}^m$. We introduce holomorphic coordinates $a_i := z_i/z_\alpha$ for $i \in \{0, \ldots, m\} \setminus \{\alpha\}$ and $b_i := z_i/z_\beta$ for $i \in \{0, \ldots, m\} \setminus \{\beta\}$ on $U_\alpha \cap U_\beta$ and we get

$$\phi_\alpha^*(\omega) = da_0 \wedge \cdots \wedge da_{\alpha-1} \wedge da_{\alpha+1} \wedge \cdots \wedge da_m,$$

$$\phi_\beta^*(\omega) = db_0 \wedge \cdots \wedge db_{\beta-1} \wedge db_{\beta+1} \wedge \cdots \wedge db_m.$$

Therefore we can write

$$db_0 \wedge \cdots \wedge db_{\beta-1} \wedge db_{\beta+1} \wedge \cdots \wedge db_m = h_{\alpha\beta} da_0 \wedge \cdots \wedge da_{\alpha-1} \wedge da_{\alpha+1} \wedge \cdots \wedge da_m.$$
$$(9.4)$$

On the other hand, $a_\beta b_\alpha = 1$ and for every $i \neq \alpha, \beta$ we have $a_i = b_i a_\beta$. This shows that $da_\beta = -(1/b_\alpha^2)db_\alpha = -a_\beta^2 db_\alpha$ and $da_i = a_\beta db_i + b_i da_\beta$ for $i \neq \alpha, \beta$. An easy algebraic computation then yields

$$da_0 \wedge \cdots \wedge da_{\alpha-1} \wedge da_{\alpha+1} \wedge \cdots \wedge da_m =$$
$$(-1)^{\alpha-\beta} a_\beta^{m+1} db_0 \wedge \cdots \wedge db_{\beta-1} \wedge db_{\beta+1} \wedge \cdots \wedge db_m.$$

Using (9.4) we thus see that the transition functions are given by

$$h_{\alpha\beta} = (-1)^{\alpha-\beta} a_\beta^{-m-1} = (-1)^{\alpha-\beta} \left(\frac{z_\alpha}{z_\beta} \right)^{m+1}.$$

Finally, denoting $c_\alpha := (-1)^\alpha$ we have $c_\alpha h_{\alpha\beta} c_\beta^{-1} = g_{\alpha\beta}^{m+1}$, which proves that

$$K_{\mathbb{C}P^m} \simeq L^{m+1}. \tag{9.5}$$

$\square$

9.4. Exercises

(1) Let $E \to M$ be a rank k complex vector bundle whose transition functions with respect to some open cover $\{U_\alpha\}$ of M are $g_{\alpha\beta}$. Show that a section $\sigma : M \to E$ of E can be identified with a collection $\{\sigma_\alpha\}$ of smooth maps $\sigma_\alpha : U_\alpha \to \mathbb{C}^k$ satisfying $\sigma_\alpha = g_{\alpha\beta}\sigma_\beta$ on $U_\alpha \cap U_\beta$.

(2) Let $\pi : E \to M$ be a complex vector bundle over a complex manifold M. Prove that E has a holomorphic structure if and only if there exists a complex structure on E as manifold, such that the projection π is a holomorphic map.

(3) Show that the tautological line bundle of $\mathbb{C}P^m$ has no non-trivial holomorphic sections.

(4) (*The hyperplane line bundle of* $\mathbb{C}P^m$) Let H denote the dual of the tautological line bundle of $\mathbb{C}P^m$. In other words, the fibre of H over some point $[z] \in \mathbb{C}P^m$ is the set of $\mathbb{C}$-linear maps $\mathbb{C}z \to \mathbb{C}$. Find local trivializations for H with holomorphic transition functions. Show that the dimension of the space of holomorphic sections of H is $m+1$.

CHAPTER 10

Hermitian bundles

10.1. The curvature operator of a connection

Let M be a smooth manifold of dimension n and let $E \to M$ be a vector bundle over M. From now on we will use the term *connection* for covariant derivative as well and we reformulate Definition 5.1 as follows:

DEFINITION 10.1. *A connection on E is a $\mathbb{C}$-linear differential operator* $\nabla : \Gamma(E) \to \Omega^1(E)$ *satisfying the Leibniz rule*

$$\nabla(f\sigma) = df \otimes \sigma + f\nabla\sigma, \quad \forall f \in C^\infty(M), \sigma \in \Gamma(E),$$

where $\Omega^1(E)$ denotes the space of E-valued 1-forms, i.e. sections of $\Lambda^1 M \otimes E$.

One can extend any connection to the bundles of E-valued p-forms on M by

$$\nabla(\omega \otimes \sigma) = d\omega \otimes \sigma + (-1)^p \omega \wedge \nabla\sigma,$$

where the wedge product has to be understood as

$$\omega \wedge \nabla\sigma := \sum_{i=1}^{n} \omega \wedge e_i^* \otimes \nabla_{e_i}\sigma$$

for any local basis $\{e_i\}$ of TM with dual basis $\{e_i^*\}$.

The *curvature operator* of ∇ is the $\mathrm{End}(E)$-valued 2-form R^∇ defined by

$$R^\nabla(\sigma) := \nabla(\nabla\sigma), \quad \forall \sigma \in \Gamma(E). \tag{10.1}$$

To check that this is indeed tensorial, we can write:

$$\nabla^2(f\sigma) = \nabla(df \otimes \sigma + f\nabla\sigma) = d^2 f \otimes \sigma - df \wedge \nabla\sigma + df \wedge \nabla\sigma + f\nabla^2\sigma = f\nabla^2\sigma.$$

More explicitly, if $\{\sigma_1, \ldots, \sigma_k\}$ are local sections of E which form a basis of each fibre over some open set U, we define the *local connection forms* $\omega_{ij} \in \Omega^1(U)$ (relative to the choice of the basis) by

$$\nabla\sigma_i = \omega_{ij} \otimes \sigma_j.$$

We also define the *local curvature 2-forms* R_{ij}^∇ by

$$R^\nabla(\sigma_i) = R_{ij}^\nabla \otimes \sigma_j,$$

and compute

$$R_{ij}^\nabla \otimes \sigma_j = R^\nabla(\sigma_i) = \nabla(\omega_{ij} \otimes \sigma_j) = (d\omega_{ij}) \otimes \sigma_j - \omega_{il} \wedge \omega_{lj} \otimes \sigma_j,$$

showing that

$$R^{\nabla}_{ij} = d\omega_{ij} - \omega_{il} \wedge \omega_{lj}. \tag{10.2}$$

The above discussion also holds if $E \to M$ is a *complex* vector bundle over M and ∇ is a $\mathbb{C}$-*linear* connection on E.

10.2. Hermitian structures and connections

Let $E \to M$ be a complex rank k bundle over some smooth manifold M.

DEFINITION 10.2. *A Hermitian structure H on E is a smooth field of Hermitian products on the fibres of E, that is, for every $x \in M$, the map $H : E_x \times E_x \to \mathbb{C}$ satisfies*

- $H(u, v)$ *is $\mathbb{C}$-linear in u for every $v \in E_x$.*
- $H(u, v) = \overline{H(v, u)}, \quad \forall\, u, v \in E_x.$
- $H(u, u) > 0, \quad \forall\, u \neq 0.$
- $H(u, v)$ *is a smooth function on M for every smooth sections u and v of E.*

A complex vector bundle endowed with a Hermitian structure is called a Hermitian vector bundle.

It is clear from the above conditions that H is $\mathbb{C}$-anti-linear in the second variable. The third condition says that H is non-degenerate. In fact, it is quite useful to view H as a $\mathbb{C}$-anti-linear isomorphism $H : E \to E^*$.

Every rank k complex vector bundle E admits Hermitian structures. To see this, just take a trivialization (U_i, ψ_i) of E and a partition of the unity (f_i) subordinate to the open cover $\{U_i\}$. For every $x \in U_i$, let $(H_i)_x$ denote the pull-back of the standard Hermitian metric on $\mathbb{C}^k$ by the $\mathbb{C}$-linear map $\psi_i|_{E_x}$. Then $H := \sum f_i H_i$ is a well-defined Hermitian structure on E.

Suppose now that M is a complex manifold. Consider the projections $\pi^{1,0} : \Lambda^1(E) \to \Lambda^{1,0}(E)$ and $\pi^{0,1} : \Lambda^1(E) \to \Lambda^{0,1}(E)$. For every connection ∇ on E, one can consider its $(1,0)$ and $(0,1)$-components $\nabla^{1,0} := \pi^{1,0} \circ \nabla$ and $\nabla^{0,1} := \pi^{0,1} \circ \nabla$. From Proposition 8.2, we can extend these operators to $\nabla^{1,0} : \Omega^{p,q}(E) \to \Omega^{p+1,q}(E)$ and $\nabla^{0,1} : \Omega^{p,q}(E) \to \Omega^{p,q+1}(E)$ satisfying the Leibniz rule:

$$\nabla^{1,0}(\omega \otimes \sigma) = \partial\omega \otimes \sigma + (-1)^{p+q}\omega \wedge \nabla^{1,0}\sigma$$

and

$$\nabla^{0,1}(\omega \otimes \sigma) = \bar\partial\omega \otimes \sigma + (-1)^{p+q}\omega \wedge \nabla^{0,1}\sigma,$$

for all $\omega \in \Omega^{p,q}M$, $\sigma \in \Gamma(E)$. Of course, $\nabla^{0,1}$ is a pseudo-holomorphic structure on E for every connection ∇.

For every section σ of E one can write

$$R^\nabla(\sigma) = \nabla^2\sigma = (\nabla^{1,0} + \nabla^{0,1})^2(\sigma)$$
$$= (\nabla^{1,0})^2(\sigma) + (\nabla^{0,1})^2(\sigma) + (\nabla^{1,0}\nabla^{0,1} + \nabla^{0,1}\nabla^{1,0})(\sigma),$$

so the $(0,2)$-type component of the curvature is given by

$$(R^\nabla)^{0,2} = (\nabla^{0,1})^2.$$

Theorem 9.2 shows that if $(R^\nabla)^{0,2}$ vanishes for some connection ∇ on E, then E is a holomorphic bundle with holomorphic structure $\bar\partial := \nabla^{0,1}$. The converse is also true: simply choose an arbitrary Hermitian metric on E and apply Theorem 10.3 below.

We say that ∇ is an H-connection (or Hermitian connection) if H, viewed as a field of $\mathbb{C}$-valued real bilinear forms on E, is parallel with respect to ∇. We can now state the main result of this section:

THEOREM 10.3. *For every Hermitian structure H in a holomorphic bundle E with holomorphic structure $\bar\partial$, there exists a unique H-connection ∇ (called the* Chern connection*) such that $\nabla^{0,1} = \bar\partial$.*

PROOF. Let us first remark that the dual vector bundle E^* is holomorphic too, with holomorphic structure still denoted by $\bar\partial$, and that any connection ∇ on E induces canonically a connection, also denoted by ∇, on E^* by the formula

$$(\nabla_X\sigma^*)(\sigma) := \partial_X(\sigma^*(\sigma)) - \sigma^*(\nabla_X\sigma), \qquad (10.3)$$

for all $X \in TM$, $\sigma \in \Gamma(E)$, $\sigma^* \in \Gamma(E^*)$. Notice that $\nabla^{0,1} = \bar\partial$ on E just means that $\nabla\sigma \in \Omega^{1,0}(E)$ for every holomorphic section σ of E. From (10.3), if this property holds on E, then it holds on E^* too.

After these preliminaries, suppose that ∇ is an H-connection with $\nabla^{0,1} = \bar\partial$. The $\mathbb{C}$-anti-linear isomorphism $H : E \to E^*$ is then ∇-parallel, so for every section σ of E and every real vector X on M we get

$$\nabla_X(H(\sigma)) = \nabla_X(H)(\sigma) + H(\nabla_X\sigma) = H(\nabla_X\sigma).$$

By the $\mathbb{C}$-anti-linearity of H, for every complex vector $Z \in TM^{\mathbb{C}}$ we have

$$\nabla_Z(H(\sigma)) = H(\nabla_{\bar Z}\sigma).$$

For $Z \in T^{1,0}M$, this shows that

$$\nabla^{1,0}(\sigma) = H^{-1} \circ \nabla^{0,1}(H(\sigma)) = H^{-1}(\bar\partial(H(\sigma))), \qquad (10.4)$$

hence $\nabla = \bar\partial + H^{-1} \circ \bar\partial \circ H$, which proves the existence and uniqueness of ∇. $\qquad\square$

REMARK. The (0,2)-component of the curvature of the Chern connection vanishes. Indeed,

$$(R^\nabla)^{0,2}(\sigma) = \nabla^{0,1}(\nabla^{0,1}(\sigma)) = \bar\partial^2(\sigma) = 0.$$

Its (2,0)-component actually vanishes too, since by (10.4),

$$\nabla^{1,0}(\nabla^{1,0}(\sigma)) = \nabla^{1,0}(H^{-1}(\bar{\partial}(H(\sigma)))) = H^{-1}(\bar{\partial}^2(H(\sigma))) = 0.$$

10.3. Exercises

(1) Let $E \to M$ be a complex vector bundle and denote by E^* and $\bar{E}$ its dual and its conjugate. Recall that for every $x \in M$, the fibre of E^* over x is just the dual of E_x and the fibre $\bar{E}_x$ of $\bar{E}$ is just E_x endowed with the conjugate complex structure, in the sense that the action of some complex number z on $\bar{E}_x$ is the same as the action of $\bar{z}$ on E_x. If $g_{\alpha\beta}$ denote the transition functions of E with respect to some open cover $\{U_\alpha\}$ of M, find the transition functions of E^* and $\bar{E}$ with respect to the same open cover.

(2) Show that a Hermitian structure on a complex vector bundle E defines an isomorphism between E^* and $\bar{E}$ as complex vector bundles.

(3) Let $E \to M$ be a rank k complex vector bundle. Viewing local trivializations of E as local bases of sections of E, show that if the transition functions of E with respect to some local trivialization take values in the unitary group $U_k \subset \mathrm{Gl}_k(\mathbb{C})$ then there exists a canonically defined Hermitian structure on E.

(4) Prove the naturality of the Chern connection with respect to direct sums and tensor products of holomorphic vector bundles.

CHAPTER 11

Hermitian and Kähler metrics

11.1. Hermitian metrics

We start with the following:

DEFINITION 11.1. *A Hermitian metric on an almost complex manifold* (M, J) *is a Riemannian metric* h *such that* $h(X, Y) = h(JX, JY)$, *for all* $X, Y \in TM$. *The* fundamental 2-form *of a Hermitian metric is defined by* $\Omega(X, Y) := h(JX, Y)$.

The extension by $\mathbb{C}$-linearity (also denoted by h) of the Hermitian metric to $TM^{\mathbb{C}}$ satisfies

$$\begin{cases} h(\bar{Z}, \bar{W}) = \overline{h(Z, W)}, & \forall \, Z, W \in TM^{\mathbb{C}}. \\ h(Z, \bar{Z}) > 0, & \text{for every non-zero complex vector } Z. \\ h(Z, W) = 0, & \forall \, Z, W \in T^{1,0}M \text{ and } \forall \, Z, W \in T^{0,1}M. \end{cases} \quad (11.1)$$

Conversely, each symmetric tensor on $TM^{\mathbb{C}}$ with these properties defines a Hermitian metric by restriction to TM (exercise).

REMARK. The tangent bundle of an almost complex manifold is in particular a complex vector bundle. If h is a Hermitian metric on M, then $H(X, Y) := h(X, Y) - ih(JX, Y) = (h - i\Omega)(X, Y)$ defines a Hermitian structure on the complex vector bundle (TM, J), as defined in the previous chapter. Conversely, any Hermitian structure H on TM as complex vector bundle defines a Hermitian metric h on M by $h := \text{Re}(H)$.

REMARK. Every almost complex manifold admits Hermitian metrics. Simply choose an arbitrary Riemannian metric g and define $h(X, Y) := g(X, Y) + g(JX, JY)$.

Let z_α be holomorphic coordinates on a complex Hermitian manifold (M^{2m}, h, J) and denote by $h_{\alpha\bar{\beta}}$ the coefficients of the metric tensor in these local coordinates:

$$h_{\alpha\bar{\beta}} := h\left(\frac{\partial}{\partial z_\alpha}, \frac{\partial}{\partial \bar{z}_\beta} \right).$$

LEMMA 11.2. *The fundamental form is given by*

$$\Omega = i \sum_{\alpha,\beta=1}^{m} h_{\alpha\bar\beta} dz_\alpha \wedge d\bar z_\beta.$$

The proof is left as an exercise.

11.2. Kähler metrics

Suppose that the fundamental form Ω of a complex Hermitian manifold is closed. The local $i\partial\bar\partial$-lemma yields the existence in some neighbourhood of each point of a real function u such that $\Omega = i\partial\bar\partial u$, which in local coordinates reads

$$h_{\alpha\bar\beta} = \frac{\partial^2 u}{\partial z_\alpha \partial \bar z_\beta}.$$

This particularly simple expression for the metric tensor in terms of one single real function deserves the following:

DEFINITION 11.3. *A Hermitian metric h on an almost complex manifold (M, J) is called a* Kähler metric *if J is a complex structure and the fundamental form Ω is closed:*

$$h \text{ is Kähler} \iff \begin{cases} N^J = 0, \\ d\Omega = 0. \end{cases}$$

A local real function u satisfying $\Omega = i\partial\bar\partial u$ is called a local Kähler potential *of the metric h.*

Our aim (as Riemannian geometers) is to express the Kähler condition in terms of the covariant derivative of the Levi–Civita connection of h. We start by doing so for the Nijenhuis tensor.

LEMMA 11.4. *Let h be a Hermitian metric on an almost complex manifold (M, J), with Levi–Civita covariant derivative ∇. Then J is integrable if and only if*

$$(\nabla_{JX} J)Y = J(\nabla_X J)Y, \qquad \forall\, X, Y \in TM. \tag{11.2}$$

PROOF. Let us fix a point $x \in M$ and extend X and Y to vector fields on M parallel with respect to ∇ at x. Then we can write

$$\begin{aligned} N^J(X,Y) &= [X,Y] + J[JX,Y] + J[X,JY] - [JX,JY] \\ &= J(\nabla_X J)Y - J(\nabla_Y J)X - (\nabla_{JX} J)Y + (\nabla_{JY} J)X \\ &= (J(\nabla_X J)Y - (\nabla_{JX} J)Y) - (J(\nabla_Y J)X - (\nabla_{JY} J)X), \end{aligned}$$

thus proving that (11.2) implies $N^J = 0$. Conversely, suppose that $N^J = 0$ and denote by $A(X, Y, Z) = h(J(\nabla_X J)Y - (\nabla_{JX} J)Y, Z)$. The previous equation just reads $A(X,Y,Z) = A(Y,X,Z)$. On the other hand, A is skew-symmetric in the last two variables, since J and $\nabla_X J$ are anti-commuting skew-symmetric operators. Thus $A(X,Y,Z) = -A(X,Z,Y)$, so by circular permutations we get

$$A(X,Y,Z) = -A(Y,Z,X) = A(Z,X,Y) = -A(X,Y,Z),$$

which implies (11.2). $\qquad\qquad\qquad\qquad\qquad\qquad\qquad\qquad\qquad\square$

THEOREM 11.5. *A Hermitian metric h on an almost complex manifold is Kähler if and only if J is parallel with respect to the Levi–Civita connection of h.*

PROOF. One direction is obvious, since if J is ∇-parallel, then N^J clearly vanishes, and as $\Omega = h(J\cdot,\cdot)$, we also have $\nabla\Omega = 0$, so by (6.5), $d\Omega = 0$. Suppose, conversely, that h is Kähler and consider the tensor $B(X,Y,Z) := h((\nabla_X J)Y, Z)$. As J and $\nabla_X J$ anti-commute we have

$$B(X,Y,JZ) = B(X,JY,Z).$$

From (11.2) we get

$$B(X,Y,JZ) + B(JX,Y,Z) = 0.$$

Combining these two relations also yields

$$B(X,JY,Z) + B(JX,Y,Z) = 0.$$

We now use $d\Omega = 0$ twice, first on X,Y,JZ, then on X,JY,Z and get:

$$B(X,Y,JZ) + B(Y,JZ,X) + B(JZ,X,Y) = 0,$$

$$B(X,JY,Z) + B(JY,Z,X) + B(Z,X,JY) = 0.$$

Adding these two relations and using the previous properties of B yields $2B(X,Y,JZ) = 0$, that is, J is ∇-parallel. $\qquad\qquad\qquad\qquad\square$

11.3. Characterization of Kähler metrics

We will now prove the analytic characterization of Kähler metrics mentioned in Section 7.1.

THEOREM 11.6. *A Hermitian metric h on a complex manifold (M,J) is Kähler if and only if around each point of M there exist holomorphic coordinates in which h osculates to the standard Hermitian metric to the order 2.*

PROOF. Suppose that for every $x \in M$ we can find holomorphic local coordinates $z_\alpha = x_\alpha + iy_\alpha$ around x such that $h_{\alpha\bar\beta} = \frac{1}{2}\delta_{\alpha\beta} + r_{\alpha\beta}$, and

$$r_{\alpha\beta}(x) = \frac{\partial r_{\alpha\beta}}{\partial x_\gamma}(x) = \frac{\partial r_{\alpha\beta}}{\partial y_\gamma}(x) = 0.$$

Then

$$d\Omega = i \sum_{\alpha,\beta,\gamma=1}^{m} \left(\frac{\partial h_{\alpha\bar\beta}}{\partial x_\gamma} dx_\gamma + \frac{\partial h_{\alpha\bar\beta}}{\partial y_\gamma} dy_\gamma \right) \wedge dz_\alpha \wedge d\bar z_\beta$$

clearly vanishes at x. As x was arbitrary, this means $d\Omega = 0$.

Conversely, if the metric is Kähler, for every $x \in M$ we take an orthonormal basis of $T_x M$ of the form $\{e_1,\ldots,e_m, Je_1,\ldots,Je_m\}$ and choose a local holomorphic coordinate system $(z_\alpha = x_\alpha + iy_\alpha)$ around x such that

$$e_\alpha = \frac{\partial}{\partial x_\alpha}(x) \quad \text{and} \quad Je_\alpha = \frac{\partial}{\partial y_\alpha}(x).$$

The Kähler 2-form Ω can be written as

$$\Omega = i \sum_{\alpha,\beta} \left(\frac{1}{2}\delta_{\alpha\beta} + \sum_\gamma a_{\alpha\beta\gamma} z_\gamma + \sum_\gamma a_{\alpha\beta\bar\gamma} \bar z_\gamma + o(|z|) \right) dz_\alpha \wedge d\bar z_\beta,$$

where $o(|z|)$ denotes generically a function whose 1-jet vanishes at x. The condition $h_{\alpha\bar\beta} = \overline{h_{\beta\bar\alpha}}$ together with Lemma 11.2 implies

$$a_{\alpha\beta\bar\gamma} = \overline{a_{\beta\alpha\gamma}}, \tag{11.3}$$

and from $d\Omega = 0$ we find

$$a_{\alpha\beta\gamma} = a_{\gamma\beta\alpha}. \tag{11.4}$$

We now look for a local coordinate change in which the Kähler form has vanishing first order terms. We put

$$z_\alpha = w_\alpha + \frac{1}{2} \sum_{\beta,\gamma} b_{\alpha\beta\gamma} w_\beta w_\gamma,$$

where $b_{\alpha\beta\gamma}$ are complex numbers satisfying $b_{\alpha\beta\gamma} = b_{\alpha\gamma\beta}$. This coordinate change is well-defined locally thanks to the holomorphic version of the local inversion theorem. We then have

$$dz_\alpha = dw_\alpha + \sum_{\beta,\gamma} b_{\alpha\beta\gamma} w_\beta dw_\gamma,$$

whence (using again the summation convention on repeating subscripts)

$$
\begin{aligned}
\Omega &= i\left(\frac{1}{2}\delta_{\alpha\beta} + a_{\alpha\beta\gamma} z_\gamma + a_{\alpha\beta\bar\gamma} \bar z_\gamma + o(|z|) \right) dz_\alpha \wedge d\bar z_\beta \\
&= i\left(\frac{1}{2}\delta_{\alpha\beta} + a_{\alpha\beta\gamma} w_\gamma + a_{\alpha\beta\bar\gamma} \bar w_\gamma + o(|w|) \right)(dw_\alpha + b_{\alpha\varepsilon\tau} w_\varepsilon dw_\tau) \\
&\qquad \wedge (d\bar w_\beta + \overline{b_{\beta\varepsilon\tau}} \bar w_\varepsilon d\bar w_\tau) \\
&= i\left(\frac{1}{2}\delta_{\alpha\beta} + a_{\alpha\beta\gamma} w_\gamma + a_{\alpha\beta\bar\gamma} \bar w_\gamma + b_{\beta\gamma\alpha} w_\gamma + \overline{b_{\alpha\gamma\beta}} \bar w_\gamma + o(|w|) \right) dw_\alpha \wedge d\bar w_\beta.
\end{aligned}
$$

If we choose $b_{\beta\gamma\alpha} = -a_{\alpha\beta\gamma}$ (which is possible because of (11.4) which ensures that $a_{\alpha\beta\gamma}$ is symmetric in α and γ), then from (11.3) we get

$$\overline{b_{\alpha\gamma\beta}} = -\overline{a_{\beta\alpha\gamma}} = -a_{\alpha\beta\bar\gamma},$$

showing that

$$\Omega = i\left(\frac{1}{2}\delta_{\alpha\beta} + o(|w|)\right)dw_\alpha \wedge d\bar w_\beta.$$

$\square$

The proof above was taken from [3], p. 107.

11.4. Comparison of the Levi–Civita and Chern connections

Our next aim is to express the $\bar\partial$-operator on the tangent bundle of a Hermitian manifold (M, h, J) in terms of the Levi–Civita connection of h. In order to do so, we have to remember that TM is identified with a complex vector bundle via the complex structure J. In other words, a product iX for some $X \in TM$ is identified with JX. Since this point is particularly confusing, we insist a little more on it: we don't say that $iX = JX$ on TM (this actually would make no sense because TM is a real bundle), we just say that the complex structure on TM (which is usually denoted by i on vector bundles) is, in this case, given by the tensor J.

LEMMA 11.7. *For every section Y of the complex vector bundle (TM, J), the $\bar\partial$-operator, as TM-valued $(0, 1)$-form is given by*

$$\bar\partial^\nabla Y(X) := \frac{1}{2}(\nabla_X Y + J\nabla_{JX}Y - J(\nabla_Y J)X), \tag{11.5}$$

where ∇ denotes the Levi–Civita connection of any Hermitian metric h on M.

PROOF. Recall first that $(\bar\partial f)(X) = \frac{1}{2}\partial_{(X+iJX)}f$, so

$$\bar\partial^\nabla(fY)(X) = f\frac{1}{2}(\nabla_X Y + J\nabla_{JX}Y - J(\nabla_Y J)X)$$

$$+ \frac{1}{2}((\partial_X f)Y + (\partial_{JX} f)JY) = f\bar\partial^\nabla Y(X) + \bar\partial f(X)Y,$$

which shows that the operator $\bar\partial^\nabla$ defined by (11.5) satisfies the Leibniz rule. Moreover, a vector field Y is a holomorphic section of TM if and only if it is real holomorphic. By Lemma 8.7, this is equivalent to $\mathcal{L}_Y J = 0$, which means that for every vector field $X \in \mathcal{X}(M)$ one has

$$\begin{aligned}
0 &= (\mathcal{L}_Y J)X = \mathcal{L}_Y(JX) - J\mathcal{L}_Y X = [Y, JX] - J[Y, X] \\
&= \nabla_Y JX - \nabla_{JX}Y - J\nabla_Y X + J\nabla_X Y \\
&= (\nabla_Y J)X - \nabla_{JX}Y + J\nabla_X Y = J(\nabla_X Y + J\nabla_{JX}Y - J(\nabla_Y J)X),
\end{aligned}$$

thus showing that $\bar{\partial}^{\nabla} Y$ vanishes for every holomorphic section Y. This proves that $\bar{\partial}^{\nabla} = \bar{\partial}$. $\qquad\square$

The tangent bundle of a Hermitian manifold (M, h, J) has two natural linear connections: the Levi–Civita connection ∇ and the Chern connection $\bar{\nabla}$.

PROPOSITION 11.8. *On a Hermitian manifold (M, h, J), the Chern connection coincides with the Levi–Civita connection if and only if (M, h, J) is Kähler.*

PROOF. Let $H := h - i\Omega$ denote the Hermitian structure of TM. By definition, J is parallel with respect to the Chern connection, which is a complex connection. Thus, if $\nabla = \bar{\nabla}$ then J is ∇-parallel, so h is Kähler by Theorem 11.5. Conversely, suppose that h is Kähler. Then $\nabla J = 0$, so the Levi–Civita connection is a complex-linear connection on TM. Moreover, it is an H-connection since $\nabla h = 0$. Finally, the condition $\nabla^{0,1} = \bar{\partial}$ follows from Lemma 11.7, as $\nabla^{0,1}_X = \frac{1}{2}(\nabla_X + i\nabla_{JX}) = \frac{1}{2}(\nabla_X + J\nabla_{JX})$. $\qquad\square$

11.5. Exercises

(1) Prove that every Hermitian metric on a 2-dimensional almost complex manifold is Kähler.

(2) Prove that the fundamental form of a Hermitian metric is a (1,1)-form.

(3) If $h_{\alpha\bar{\beta}}$ denote the coefficients of a Hermitian metric tensor in some local holomorphic coordinate system, show that $h_{\alpha\bar{\beta}} = \overline{h_{\beta\bar{\alpha}}}$.

(4) Show that the extension of a Hermitian metric h by $\mathbb{C}$-linearity is a symmetric bilinear tensor satisfying
$$\begin{cases} h(\bar{Z}, \bar{W}) = \overline{h(Z, W)}, & \forall\, Z, W \in TM^{\mathbb{C}}, \\ h(Z, \bar{Z}) > 0, & \text{for every non-zero complex vector } Z, \\ h(Z, W) = 0, & \forall\, Z, W \in T^{1,0}M \text{ and } \forall\, Z, W \in T^{0,1}M, \end{cases}$$
and conversely, any symmetric complex bilinear tensor satisfying this system arises from a Hermitian metric.

CHAPTER 12

The curvature tensor of Kähler manifolds

12.1. The Kählerian curvature tensor

Let (M^{2m}, h, J) be a Kähler manifold with Levi–Civita covariant derivative ∇. We denote as before by R its curvature tensor (which satisfies the symmetries given by Lemma 6.6) and by Ric its Ricci tensor

$$\mathrm{Ric}(X, Y) := \mathrm{Tr}\{V \mapsto R(V, X)Y\}.$$

Since J is ∇-parallel, the Riemannian curvature tensor satisfies:

$$R(X, Y)JZ = JR(X, Y)Z, \quad \forall\, X, Y, Z \in \mathcal{X}(M). \tag{12.1}$$

By Lemma 6.6 this immediately implies

$$R(X, Y, JZ, JT) = R(X, Y, Z, T) = R(JX, JY, Z, T),$$

whence, in a local orthonormal basis $\{e_i\}$,

$$\mathrm{Ric}(JX, JY) = \sum_{i=1}^{2m} R(e_i, JX, JY, e_i) = \sum_{i=1}^{2m} R(Je_i, X, Y, Je_i) = \mathrm{Ric}(X, Y),$$

since the set $\{Je_i\}$ is also an orthonormal basis. This last equation shows that the expression $\mathrm{Ric}(JX, Y)$ is skew-symmetric in X and Y, therefore justifying the following:

DEFINITION 12.1. *The* Ricci form ρ *of a Kähler manifold is defined by*

$$\rho(X, Y) := \mathrm{Ric}(JX, Y), \quad \forall\, X, Y \in TM.$$

The Ricci form is one of the most important objects on a Kähler manifold. Among its properties which will be proved in the next chapters we mention:

- the Ricci form ρ is closed;
- the cohomology class of ρ is equal (up to some real multiple) to the Chern class of the canonical bundle of M;
- in local coordinates, ρ can be expressed as $\rho = -i\partial\bar\partial \log \det(h_{\alpha\bar\beta})$, where $\det(h_{\alpha\bar\beta})$ denotes the determinant of the matrix $(h_{\alpha\bar\beta})$ expressing the Hermitian metric.

For the moment, we use the Bianchi identities satisfied by the curvature tensor to prove the following:

Proposition 12.2. (i) *The Ricci tensor of a Kähler manifold satisfies*

$$\mathrm{Ric}(X, Y) = \frac{1}{2}\mathrm{Tr}(R(X, JY) \circ J).$$

(ii) *The Ricci form is closed.*

Proof. (i) Let (e_i) be a local orthonormal basis of TM. Using the first Bianchi identity we get

$$
\begin{aligned}
\mathrm{Ric}(X, Y) &= \sum_{i=1}^{2m} R(e_i, X, Y, e_i) = \sum_{i=1}^{2m} R(e_i, X, JY, Je_i) \\
&= \sum_{i=1}^{2m}(-R(X, JY, e_i, Je_i) - R(JY, e_i, X, Je_i)) \\
&= \sum_{i=1}^{2m}(R(X, JY, Je_i, e_i) + R(Y, Je_i, X, Je_i)) \\
&= \mathrm{Tr}(R(X, JY) \circ J) - \mathrm{Ric}(X, Y).
\end{aligned}
$$

(ii) From (i) we can write $2\rho(X, Y) = \mathrm{Tr}(R(X, Y) \circ J)$. Therefore,

$$
\begin{aligned}
2d\rho(X, Y, Z) \ &\overset{(6.5)}{=}\ 2((\nabla_X \rho)(Y, Z) + (\nabla_Y \rho)(Z, X) + (\nabla_Z \rho)(X, Y)) \\
&= \mathrm{Tr}\big(((\nabla_X R)(Y, Z) + (\nabla_Y R)(Z, X) + (\nabla_Z R)(X, Y)) \circ J\big) \\
&= 0,
\end{aligned}
$$

the last equality being a direct consequence of the second Bianchi identity.   □

12.2. The curvature tensor in local coordinates

Let (M^{2m}, h, J) be a Kähler manifold with Levi–Civita covariant derivative ∇ and let (z_α) be a system of local holomorphic coordinates. We consider the associate local basis of the complexified tangent bundle $T^{\mathbb{C}}M$:

$$Z_\alpha := \frac{\partial}{\partial z_\alpha}, \qquad Z_{\bar\alpha} := \frac{\partial}{\partial \bar z_\alpha}, \qquad 1 \le \alpha \le m.$$

We let Roman subscripts $A, B, C, \ldots$ run over the set $\{1, \ldots, m, \bar 1, \ldots, \bar m\}$ and Greek subscripts $\alpha, \beta, \gamma, \ldots$ run over the set $\{1, \ldots, m\}$. We denote the components of the Kähler metric in the above coordinates by

$$h_{AB} := h(Z_A, Z_B).$$

Of course, since the metric is Hermitian we have

$$h_{\alpha\beta} = h_{\bar\alpha\bar\beta} = 0, \qquad h_{\bar\beta\alpha} = h_{\alpha\bar\beta} = \overline{h_{\bar\beta\bar\alpha}}. \tag{12.2}$$

Let $h^{\alpha\bar{\beta}}$ denote the coefficients of the inverse matrix of $(h_{\alpha\bar{\beta}})$. Using, as usual, the summation convention on repeating subscripts, we define the *Christoffel symbols* relative to the basis $\{Z_A\}$ of $TM^{\mathbb{C}}$ by

$$\nabla_{Z_A} Z_B = \Gamma_{AB}^C Z_C.$$

Using the convention $\bar{\bar{\alpha}} = \alpha$, we get by conjugation

$$\overline{\Gamma_{AB}^C} = \Gamma_{\bar{A}\bar{B}}^{\bar{C}}.$$

Since ∇ is torsion-free we have

$$\Gamma_{AB}^C = \Gamma_{BA}^C,$$

and since $T^{1,0}$ is ∇-parallel we must have

$$\Gamma_{A\bar{\beta}}^{\gamma} = 0.$$

These relations show that the only non-vanishing Christoffel symbols are

$$\Gamma_{\alpha\beta}^{\gamma} \qquad \text{and} \qquad \Gamma_{\bar{\alpha}\bar{\beta}}^{\bar{\gamma}}.$$

In order to compute them, we notice that $\Gamma_{\alpha\bar{\delta}}^C = 0$ implies

$$\nabla_{Z_\alpha} Z_{\bar{\delta}} = 0, \tag{12.3}$$

hence

$$\frac{\partial h_{\beta\bar{\delta}}}{\partial z_\alpha} = h(\nabla_{Z_\alpha} Z_\beta, Z_{\bar{\delta}}) = \Gamma_{\alpha\beta}^{\gamma} h_{\gamma\bar{\delta}}.$$

This proves the formulas

$$\Gamma_{\alpha\beta}^{\gamma} h_{\gamma\bar{\delta}} = \frac{\partial h_{\beta\bar{\delta}}}{\partial z_\alpha} \qquad \text{and} \qquad \Gamma_{\alpha\beta}^{\gamma} = h^{\gamma\bar{\delta}} \frac{\partial h_{\beta\bar{\delta}}}{\partial z_\alpha}. \tag{12.4}$$

The curvature tensor can be viewed either as $(3,1)$- or as $(4,0)$-tensor. The corresponding coefficients are defined by

$$R(Z_A, Z_B) Z_C = R_{ABC}^D Z_D$$

and

$$R_{ABCD} = R(Z_A, Z_B, Z_C, Z_D) = h_{DE} R_{ABC}^E.$$

From the fact that $T^{1,0} M$ is parallel we immediately get $R_{AB\bar{\delta}}^{\gamma} = R_{AB\delta}^{\bar{\gamma}} = 0$, hence $R_{AB\gamma\delta} = R_{AB\bar{\gamma}\bar{\delta}} = 0$. Using the curvature symmetries we finally see that the only non-vanishing components of R are

$$R_{\alpha\bar{\beta}\gamma\bar{\delta}}, \; R_{\alpha\bar{\beta}\bar{\gamma}\delta}, \; R_{\bar{\alpha}\beta\gamma\bar{\delta}}, \; R_{\bar{\alpha}\beta\bar{\gamma}\delta}$$

and

$$R_{\alpha\bar{\beta}\gamma}^{\delta}, \; R_{\alpha\bar{\beta}\bar{\gamma}}^{\bar{\delta}}, \; R_{\bar{\alpha}\beta\gamma}^{\delta}, \; R_{\bar{\alpha}\beta\bar{\gamma}}^{\bar{\delta}}.$$

From (12.3) and (12.4) we obtain

$$R_{\alpha\bar{\beta}\gamma}^{\delta} Z_\delta = R(Z_\alpha, Z_{\bar{\beta}}) Z_\gamma = -\nabla_{Z_{\bar{\beta}}}(\nabla_{Z_\alpha} Z_\gamma) = -\nabla_{Z_{\bar{\beta}}}(\Gamma_{\alpha\gamma}^{\delta} Z_\delta) = -\frac{\partial \Gamma_{\alpha\gamma}^{\delta}}{\partial z_{\bar{\beta}}} Z_\delta,$$

therefore

$$R^{\delta}_{\alpha\bar{\beta}\gamma} = -\frac{\partial\Gamma^{\delta}_{\alpha\gamma}}{\partial\bar{z}_{\beta}}. \tag{12.5}$$

Using this formula we can compute the components of the Ricci tensor:

$$\text{Ric}_{\gamma\bar{\beta}} = \text{Ric}_{\bar{\beta}\gamma} = R^{A}_{A\bar{\beta}\gamma} = R^{\alpha}_{\alpha\bar{\beta}\gamma} = -\frac{\partial\Gamma^{\alpha}_{\alpha\gamma}}{\partial\bar{z}_{\beta}}.$$

Let us denote by d the determinant of the matrix $(h_{\alpha\bar{\beta}})$. From Lemma 12.3 below and (12.4) we get

$$\Gamma^{\alpha}_{\alpha\gamma} = \Gamma^{\alpha}_{\gamma\alpha} = h^{\alpha\bar{\delta}}\frac{\partial h_{\alpha\bar{\delta}}}{\partial z_{\gamma}} = \frac{1}{d}\frac{\partial d}{\partial z_{\gamma}} = \frac{\partial\log d}{\partial z_{\gamma}}.$$

This proves the following simple expressions for the Ricci tensor

$$\text{Ric}_{\alpha\bar{\beta}} = -\frac{\partial^{2}\log d}{\partial z_{\alpha}\partial\bar{z}_{\beta}},$$

and for the Ricci form

$$\rho = -i\partial\bar{\partial}\log d. \tag{12.6}$$

LEMMA 12.3. *Let* $(h_{ij}) = (h_{ij}(t))$ *be the coefficients of a map* $h : \mathbb{R} \to \text{Gl}_m(\mathbb{C})$ *with* $h^{ij} := (h_{ij})^{-1}$ *and let* $d(t)$ *denote the determinant of* (h_{ij}). *Then the following formula holds*

$$d'(t) = d\sum_{i,j=1}^{m} h'_{ij}(t)h^{ji}(t).$$

PROOF. Recall the definition of the determinant

$$d = \sum_{\sigma\in\mathfrak{S}_m} \varepsilon(\sigma)h_{1\sigma_1}\ldots h_{m\sigma_m}.$$

If we denote

$$\tilde{h}^{ji} := \frac{1}{d}\sum_{\substack{\sigma\in\mathfrak{S}_m \\ \sigma_i=j}} \varepsilon(\sigma)h_{1\sigma_1}\ldots h_{i-1\sigma_{i-1}}h_{i+1\sigma_{i+1}}\ldots h_{m\sigma_m},$$

then we obtain easily

$$\sum_{j=1}^{m} h_{ij}\tilde{h}^{ji} = \frac{1}{d}\sum_{j=1}^{m}\sum_{\substack{\sigma\in\mathfrak{S}_m \\ \sigma_i=j}} \varepsilon(\sigma)h_{1\sigma_1}\ldots h_{m\sigma_m} = \frac{1}{d}\sum_{\sigma\in\mathfrak{S}_m} \varepsilon(\sigma)h_{1\sigma_1}\ldots h_{m\sigma_m}$$

$$= \frac{1}{d}d = 1,$$

and

$$\sum_{j=1}^{m} h_{kj}\tilde{h}^{ji} = \frac{1}{d}\sum_{j=1}^{m}\sum_{\substack{\sigma\in\mathfrak{S}_m \\ \sigma_i=j}} \varepsilon(\sigma)h_{1\sigma_1}\ldots h_{i-1\sigma_{i-1}}h_{k\sigma_i}h_{i+1\sigma_{i+1}}\ldots h_{m\sigma_m} = 0$$

for $k \neq i$ since in the last sum each term corresponding to a permutation σ is the opposite of the term corresponding to the permutation $(ik) \circ \sigma$, where (ik) denotes the transposition of i and k. This shows that $\tilde{h}^{ji} = h^{ji}$ are the coefficients of the inverse matrix of h. We now get

$$
\begin{aligned}
d'(t) &= \sum_{\sigma \in \mathfrak{S}_m} \sum_{i=1}^{m} \varepsilon(\sigma) h'_{i\sigma_i}(t) h_{1\sigma_1} \ldots h_{i-1\sigma_{i-1}} h_{i+1\sigma_{i+1}} \ldots h_{m\sigma_m} \\
&= \sum_{i=1}^{m} \sum_{j=1}^{m} \sum_{\substack{\sigma \in \mathfrak{S}_m \\ \sigma_i = j}} \varepsilon(\sigma) h'_{ij}(t) h_{1\sigma_1} \ldots h_{i-1\sigma_{i-1}} h_{i+1\sigma_{i+1}} \ldots h_{m\sigma_m} \\
&= d \sum_{i,j=1}^{m} h'_{ij}(t) \tilde{h}^{ji} = d \sum_{i,j=1}^{m} h'_{ij}(t) h^{ji}(t).
\end{aligned}
$$

$\square$

12.3. Exercises

(1) Let $S := \mathrm{Tr}(\mathrm{Ric})$ denote the scalar curvature of a Kähler manifold M with Ricci form ρ. Using the second Bianchi identity, prove the formula:

$$
\delta\rho = -\frac{1}{2} J dS, \tag{12.7}
$$

where δ denotes the codifferential (see (14.7) below).

(2) Prove that the curvature of a Kähler manifold M, viewed as a symmetric endomorphism of the space of complex 2-forms, maps $\Lambda^{0,2}M$ and $\Lambda^{2,0}M$ to 0. Compute the image of the Kähler form through this endomorphism.

(3) Let h be a Hermitian metric on some complex manifold M^{2m} and let $z_\alpha = x_\alpha + iy_\alpha$ be a local system of holomorphic local coordinates on M. Using (12.2), show that the determinant of the complex $m \times m$ matrix $(h_{\alpha\bar{\beta}})$ is a positive real number whose square is equal to the determinant of the real $2m \times 2m$ matrix h_{ij} representing the metric in the local coordinate system (x_i, y_i).

(4) Let h and h' be two Kähler metrics on some complex manifold (M, J) having the same (Riemannian) volume form. Prove that the Ricci tensors of h and h' are equal.

CHAPTER 13

Examples of Kähler metrics

13.1. The flat metric on $\mathbb{C}^m$

Its coefficients with respect to the canonical holomorphic coordinates are

$$h_{\alpha\bar\beta} = h\left(\frac{\partial}{\partial z_\alpha}, \frac{\partial}{\partial \bar z_\beta}\right) = \frac{1}{4}h\left(\frac{\partial}{\partial x_\alpha} - i\frac{\partial}{\partial y_\alpha}, \frac{\partial}{\partial x_\beta} + i\frac{\partial}{\partial y_\beta}\right) = \frac{1}{2}\delta_{\alpha\beta},$$

so by Lemma 11.2, the Kähler form is given by

$$\Omega = i\frac{1}{2}\sum_{\alpha=1}^{m} dz_\alpha \wedge d\bar z_\alpha = \frac{i}{2}\partial\bar\partial|z|^2.$$

Thus $u(z) = \frac{1}{2}|z|^2$ is a Kähler potential for the canonical Hermitian metric on $\mathbb{C}^m$.

13.2. The Fubini–Study metric on the complex projective space

Consider the canonical holomorphic atlas (U_j, ϕ_j) on $\mathbb{C}P^m$ described in Section 7.3. Let $\pi : \mathbb{C}^{m+1} \setminus \{0\} \to \mathbb{C}P^m$ be the canonical projection

$$\pi(z_0, \ldots, z_m) = [z_0 : \ldots : z_m].$$

This map is clearly onto. It is moreover a principal $\mathbb{C}^*$-fibration, with local trivializations $\psi_j : \pi^{-1}U_j \to U_j \times \mathbb{C}^*$ given by

$$\psi_j(z) = ([z], z_j),$$

and transition functions $\psi_j \circ \psi_k^{-1}([z], \alpha) = ([z], (z_j/z_k)\alpha)$.

Consider the functions $u : \mathbb{C}^m \to \mathbb{R}$ and $v : \mathbb{C}^{m+1} \setminus \{0\} \to \mathbb{R}$ defined by $u(w) = \log(1+|w|^2)$ and $v(z) = \log(|z|^2)$. For every $j \in \{0, \ldots, m\}$, we define $f_j = \phi_j \circ \pi$.

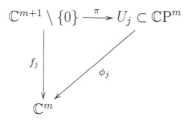

The map f_j is clearly holomorphic and a direct calculation yields

$$u \circ f_j(z) = v(z) - \log |z_j|^2.$$

As $\partial\bar\partial \log |z_j|^2 = 0$, this shows that $(f_j)^*(\partial\bar\partial u) = \partial\bar\partial v$ for every j. We thus can define a global 2-form Ω on $\mathbb{C}P^m$ by

$$\Omega|_{U_j} := i(\phi_j)^*(\partial\bar\partial u),$$

which satisfies

$$\pi^*(\Omega) = i\partial\bar\partial v. \tag{13.1}$$

Clearly Ω is a closed real $(1,1)$-form, so the tensor h defined by

$$h(X, Y) := \Omega(X, JY), \quad \forall\, X, Y \in T\mathbb{C}P^m$$

is symmetric and Hermitian. The next lemma proves that h defines a Kähler metric on $\mathbb{C}P^m$.

LEMMA 13.1. *The tensor h is positive definite on $\mathbb{C}P^m$.*

PROOF. Consider a canonical holomorphic chart $\phi_j : U_j \to \mathbb{C}^m$. Clearly, $h = (\phi_j)^*(\hat h)$, where $\hat h$ is the symmetric tensor on $\mathbb{C}^m$ defined by $\hat h(X, Y) := i\partial\bar\partial u(X, JY)$, $\forall\, X, Y \in T\mathbb{C}^m$. We have to prove that $\hat h$ is positive definite. Now, since the unitary group U_m consists of holomorphic transformations of $\mathbb{C}^m$ preserving the function u, it also preserves $\hat h$. Moreover, U_m acts transitively on the unit sphere of $\mathbb{C}^m$, so it is enough to prove that $\hat h$ is positive definite at a point $p = (r, 0, \ldots, 0) \in \mathbb{C}^m$ for some positive real number r. We have

$$\partial\bar\partial \log(1 + |z|^2) = \partial\left(\frac{1}{1+|z|^2} \left(\sum_{i=1}^{m} z_i d\bar z_i \right) \right) = \frac{1}{1+|z|^2} \sum_{i=1}^{m} dz_i \wedge d\bar z_i$$

$$- \frac{1}{(1+|z|^2)^2} \left(\sum_{i=1}^{m} \bar z_i dz_i \right) \wedge \left(\sum_{i=1}^{m} z_i d\bar z_i \right).$$

At p this 2-form simplifies to

$$\frac{1}{(1+r^2)^2} \left(dz_1 \wedge d\bar z_1 + (1+r^2) \sum_{i=2}^{m} dz_i \wedge d\bar z_i \right),$$

which shows that

$$\hat h_p(X, Y) = \frac{2}{(1+r^2)^2} \mathrm{Re}\left(X_1 \bar Y_1 + (1+r^2) \sum_{i=2}^{m} X_i \bar Y_i \right)$$

is positive definite. $\qquad\square$

The Kähler metric on $\mathbb{C}P^m$ constructed in this way is called the *Fubini–Study metric* and is usually denoted by h_{FS}.

13.3. Geometrical properties of the Fubini–Study metric

The Fubini–Study metric was defined via its Kähler 2-form, which was expressed by local Kähler potentials. We provide here a more geometrical description of this metric, showing that it is the projection onto $\mathbb{C}P^m$ of some symmetric tensor field of $\mathbb{C}^{m+1} \setminus \{0\}$.

LEMMA 13.2. *The canonical projection* $\pi : \mathbb{C}^{m+1} \setminus \{0\} \to \mathbb{C}P^m$ *is a submersion, and for every* $z \in \mathbb{C}^{m+1} \setminus \{0\}$, *the kernel of its differential at* z, $d\pi_z : T_z(\mathbb{C}^{m+1} \setminus \{0\}) \to T_{\pi(z)}\mathbb{C}P^m$, *is the complex line spanned by* z.

PROOF. Let $z \in \mathbb{C}^{m+1}$ with $z_j \neq 0$. The composition $f_j := \phi_j \circ \pi$ is given by

$$f_j(z_0, \ldots, z_m) = \frac{1}{z_j}(z_0, \ldots, z_{j-1}, z_{j+1}, \ldots, z_m).$$

We take $j = 0$ for simplicity and denote $f = f_0$. Its differential at z applied to some tangent vector v is

$$df_z(v) = \frac{1}{z_0}(v_1, \ldots, v_m) - \frac{v_0}{z_0^2}(z_1, \ldots, z_m).$$

Thus $v \in \ker(d\pi_z) \iff v \in \ker(df_z) \iff v = (v_0/z_0)z$. This shows that $\ker(d\pi_z)$ is the complex line spanned by z, and for dimensional reasons $d\pi_z$ has to be onto. $\qquad\square$

Consider the complex orthogonal $z^\perp$ of z in $\mathbb{C}^{m+1}$ with respect to the canonical Hermitian metric, i.e. the set

$$z^\perp := \{y \in \mathbb{C}^{m+1} \mid \sum_{j=0}^{m} z_j \bar{y}_j = 0\}.$$

Taking $D_z := z^\perp$ for all $z \in \mathbb{C}^{m+1} \setminus \{0\}$, defines a codimension 1 complex distribution D of the tangent bundle of $\mathbb{C}^{m+1} \setminus \{0\}$. Let $X \mapsto X^\perp$ denote the orthogonal projection onto $z^\perp$ in $T_z(\mathbb{C}^{m+1} \setminus \{0\})$ and define a bilinear symmetric tensor $\tilde{h}$ on $\mathbb{C}^{m+1} \setminus \{0\}$ by

$$\tilde{h}_z(X, Y) := \frac{2}{|z|^2}\langle X^\perp, Y^\perp \rangle, \qquad \forall\, X, Y \in T_z(\mathbb{C}^{m+1} \setminus \{0\}),$$

where $\langle \cdot, \cdot \rangle$ denotes the canonical Hermitian product.

LEMMA 13.3. *The* $(1,1)$-*form* $\varphi(X, Y) := \tilde{h}(JX, Y)$ *associated to the tensor* $\tilde{h}$ *satisfies* $\varphi = i\partial\bar{\partial} \log(|z|^2)$ *on* $\mathbb{C}^{m+1} \setminus \{0\}$.

PROOF. It is enough to prove this relation at a point $p = (r, 0, \ldots, 0) \in \mathbb{C}^{m+1} \setminus \{0\}$ for some positive real number r because both members are invariant under the action of the unitary group U_{m+1}, which is transitive on spheres. We have

$$\partial\bar{\partial}\log(|z|^2) = \partial\left(\frac{1}{|z|^2}\left(\sum_{i=0}^{m} z_i d\bar{z}_i\right)\right)$$

$$= \frac{1}{|z|^2}\sum_{i=0}^{m} dz_i \wedge d\bar{z}_i - \frac{1}{|z|^4}\left(\sum_{i=0}^{m} \bar{z}_i dz_i\right) \wedge \left(\sum_{i=0}^{m} z_i d\bar{z}_i\right).$$

At p, this 2-form simplifies to

$$(\partial\bar{\partial}\log(|z|^2))_p = \frac{1}{r^2}\sum_{i=1}^{m}(dz_i \wedge d\bar{z}_i)_p.$$

On the other hand, we have at p

$$-i\varphi\left(\frac{\partial}{\partial z_\alpha}, \frac{\partial}{\partial \bar{z}_\beta}\right) = -i\tilde{h}\left(i\frac{\partial}{\partial z_\alpha}, \frac{\partial}{\partial \bar{z}_\beta}\right) = \tilde{h}\left(\frac{\partial}{\partial z_\alpha}, \frac{\partial}{\partial \bar{z}_\beta}\right)$$

which vanishes if $\alpha = 0$ or $\beta = 0$ and equals $(1/r^2)\delta_{\alpha\beta}$ otherwise. Thus

$$-i\varphi_p = \frac{1}{r^2}\sum_{i=1}^{m}(dz_i \wedge d\bar{z}_i)_p = (\partial\bar{\partial}\log(|z|^2))_p.$$

$\square$

Using this lemma and (13.1) we see that $\pi^*h = \tilde{h}$, showing that the Fubini–Study metric h_{FS} on $\mathbb{CP}^m$ is given by the projection of the above defined semi-positive symmetric tensor field $\tilde{h}$.

PROPOSITION 13.4. *The unitary group* U_{m+1} *acts transitively by holomorphic isometries on* $(\mathbb{CP}^m, h_{FS})$.

PROOF. For every $A \in \mathrm{U}_{m+1}$, $z \in \mathbb{C}^{m+1} \setminus \{0\}$ and $\alpha \in \mathbb{C}^*$, we have $A(\alpha z) = \alpha A(z)$, showing that the canonical action of U_{m+1} on $\mathbb{C}^{m+1} \setminus \{0\}$ descends to an action on $\mathbb{CP}^m$. For every $A \in \mathrm{U}_{m+1}$, let $\tilde{A}$ be the corresponding transformation of $\mathbb{CP}^m$. Its expression in the canonical holomorphic charts shows that every $\tilde{A}$ acts *holomorphically* on $\mathbb{CP}^m$. In order to check that $\tilde{A}$ preserves the Fubini–Study metric, we first use (13.1) and the relation $v \circ A(z) = \log|Az|^2 = \log|z|^2 = v(z)$ to get

$$\pi^*(\tilde{A}^*(\Omega)) = A^*(i\partial\bar{\partial}v) = i\partial\bar{\partial}A^*v = i\partial\bar{\partial}v = \pi^*\Omega.$$

Lemma 13.2 shows that π_* is onto, so π^* is injective on exterior forms, hence $\tilde{A}^*(\Omega) = \Omega$. Since $\tilde{A}$ preserves the complex structure too, this clearly implies that $\tilde{A}$ is an isometry. $\square$

We will now use the computations in local coordinates performed in Section 12.2 in order to show that the Fubini–Study metric is Einstein. Since there exists a transitive isometric action on $\mathbb{CP}^m$, it is enough to check this at some point, say $p := [1 : 0 : \ldots : 0] \in \mathbb{CP}^m$. From Lemma 13.1 we see that the Kähler form is given in the local chart ϕ_0 by

$$(\phi_0^*)^{-1}\Omega = \frac{i}{1 + |z|^2} \sum_{i=1}^{m} dz_i \wedge d\bar{z}_i - \frac{i}{(1 + |z|^2)^2} \left(\sum_{i=1}^{m} \bar{z}_i dz_i \right) \wedge \left(\sum_{i=1}^{m} z_i d\bar{z}_i \right).$$

LEMMA 13.5. *Let dx denote the volume form on $\mathbb{C}^m$*

$$dx := dx_1 \wedge dy_1 \wedge \cdots \wedge dx_m \wedge dy_m = \frac{i}{2} dz_1 \wedge d\bar{z}_1 \wedge \cdots \wedge \frac{i}{2} dz_m \wedge d\bar{z}_m.$$

Then the local expression of the Kähler 2-form Ω in the chart ϕ_0 satisfies

$$((\phi_0^*)^{-1}\Omega)^m = \frac{2^m m!}{(1 + |z|^2)^{m+1}} dx.$$

PROOF. Both terms are clearly invariant by the action of U_m on $\mathbb{C}^m$, which is transitive on spheres, so it is enough to prove the equality at points of the form $z = (r, 0, \ldots, 0)$, where it is actually obvious. □

Now, for every Hermitian metric h on $\mathbb{C}^m$ with fundamental form φ, the determinant d of the matrix $(h_{\alpha\bar{\beta}})$ satisfies

$$\frac{1}{m!}\varphi^m = d2^m dx.$$

Applying this to our situation and using the lemma above yields

$$d = \det(h_{\alpha\bar{\beta}}) = \frac{1}{(1 + |z|^2)^{m+1}},$$

whence $\log d = -(m + 1) \log(1 + |z|^2)$, so from the local formula (12.6) for the Ricci form we get

$$\rho = -i\partial\bar{\partial} \log d = (m + 1)i\partial\bar{\partial} \log(1 + |z|^2) = (m + 1)\Omega,$$

thus proving that the Fubini–Study metric on $\mathbb{CP}^m$ is an Einstein metric, with Einstein constant $m + 1$.

13.4. Exercises

(1) A submersion $f : (M, g) \to (N, h)$ between Riemannian manifolds is called *Riemannian submersion* if for every $x \in M$, the restriction of $(f_*)_x$ to the g-orthogonal of the tangent space to the fibre $f^{-1}(f(x))$ is an isometry onto $T_{f(x)}N$. Prove that the restriction of the canonical projection π to S^{2m+1} defines a Riemannian submersion onto $(\mathbb{CP}^m, \frac{1}{2}h_{FS})$.

(2) Show that $(\mathbb{CP}^1, h_{FS})$ is isometric to the round sphere of radius $1/\sqrt{2}$, $S^2(1/\sqrt{2}) \subset \mathbb{R}^3$. *Hint:* Use the fact that a simply-connected manifold with constant positive sectional curvature K is isometric to the sphere of radius $1/\sqrt{K}$.

(3) Show that for every Hermitian tensor h on $\mathbb{C}^m$ with fundamental form φ, the determinant d of the matrix $(h_{\alpha\bar{\beta}})$ satisfies

$$\varphi^m = d2^m m! dx.$$

(4) Let Ω be the fundamental 2-form of an almost Hermitian manifold (M^{2m}, h, J). Show that the Riemannian volume form dv of h satisfies $m! dv = \Omega^m$.

CHAPTER 14

Natural operators on Riemannian and Kähler manifolds

14.1. The formal adjoint of a linear differential operator

Let (M^n, g) be an oriented Riemannian manifold (not necessarily compact) with volume form dv and let E and F be Hermitian vector bundles over M with Hermitian structures denoted by $\langle \cdot, \cdot \rangle_E$ and $\langle \cdot, \cdot \rangle_F$.

DEFINITION 14.1. *Let $P : \Gamma(E) \to \Gamma(F)$ and $Q : \Gamma(F) \to \Gamma(E)$ be linear differential operators. The operator Q is called a* formal adjoint *of P if*

$$\int_M \langle P\alpha, \beta \rangle_F dv = \int_M \langle \alpha, Q\beta \rangle_E dv,$$

for every compactly supported smooth sections $\alpha \in C_0^\infty(E)$ and $\beta \in C_0^\infty(F)$.

LEMMA 14.2. *There exists at most one formal adjoint for every linear differential operator.*

PROOF. Suppose that $P : \Gamma(E) \to \Gamma(F)$ has two formal adjoints, denoted Q and Q'. Then their difference $R := Q - Q'$ satisfies

$$\int_M \langle \alpha, R\beta \rangle_E dv = 0, \qquad \forall\, \alpha \in C_0^\infty(E), \ \forall\, \beta \in C_0^\infty(F). \tag{14.1}$$

Suppose that there exists some $\sigma \in \Gamma(F)$ and some $x \in M$ such that $R(\sigma)(x) \neq 0$. Take a positive bump function f on M such that $f \equiv 1$ on some open set U containing x and $f = 0$ outside a compact set (see Section 1.3). Since R is a differential operator, the value of $R(\sigma)$ at x only depends on the germ of σ at x, so in particular $R(f\sigma)$ has compact support and $R(f\sigma)(x) = R(\sigma)(x) \neq 0$. Applying the formula (14.1) above to the compactly supported sections $\alpha := R(f\sigma)$ and $\beta := f\sigma$ of E and F we get

$$0 = \int_M \langle \alpha, R\beta \rangle_E dv = \int_M |R(f\sigma)|^2 dv.$$

This shows that the smooth positive function $|R(f\sigma)|^2$ has to vanish identically on M, contradicting the fact that its value at x is non-zero. $\qquad \square$

The formal adjoint of an operator P is usually denoted by P^*. From the above lemma it can be checked immediately that P is the formal adjoint of

P^* and that $Q^* \circ P^*$ is the formal adjoint of $P \circ Q$. The lemma below gives a useful method to compute the formal adjoint:

LEMMA 14.3. *Let* $P : \Gamma(E) \to \Gamma(F)$ *and* $Q : \Gamma(F) \to \Gamma(E)$ *be linear differential operators. If there exists a section* $\omega \in \Gamma(E^* \otimes F^* \otimes \Lambda^{n-1}M)$ *such that*

$$(\langle P\alpha, \beta \rangle_F - \langle \alpha, Q\beta \rangle_E)dv = d(\omega(\alpha, \beta)), \quad \forall\, \alpha \in \Gamma(E),\ \forall\, \beta \in \Gamma(F), \quad (14.2)$$

then Q *is the formal adjoint of* P.

PROOF. For every compactly supported sections α and β of E and F respectively, the $(n-1)$-form $\omega(\alpha, \beta)$ has compact support. By the Stokes theorem, the integral over M of its exterior derivative vanishes. □

14.2. The Laplace operator on Riemannian manifolds

We consider an oriented Riemannian manifold (M^n, g) with volume form dv. We denote generically by $\{e_1, \dots, e_n\}$ a local orthonormal frame on M parallel at a point (Lemma 6.3) and identify vectors and 1-forms via the metric g. In this way we can write for instance $dv = e_1 \wedge \cdots \wedge e_n$.

There is a natural embedding φ of $\Lambda^k M$ in $(T^*M)^{\otimes k}$ given by

$$\varphi(\omega)(X_1, \dots, X_k) := \omega(X_1, \dots, X_k),$$

which in the above local basis reads

$$\varphi(e_1 \wedge \cdots \wedge e_k) = \sum_{\sigma \in \mathfrak{S}_k} \varepsilon(\sigma) e_{\sigma_1} \otimes \cdots \otimes e_{\sigma_k}.$$

The Riemannian product g induces a Riemannian product on all tensor bundles. We consider the following weighted scalar product on $\Lambda^k M$:

$$\langle \omega, \tau \rangle := \frac{1}{k!} g(\varphi(\omega), \varphi(\tau)),$$

which can also be characterized by the fact that the basis

$$\{e_{i_1} \wedge \cdots \wedge e_{i_k} \mid 1 \le i_1 < \cdots < i_k \le n\}$$

is orthonormal. With respect to this scalar product, the interior and exterior products are adjoint operators:

$$\langle X \lrcorner \omega, \tau \rangle = \langle \omega, X \wedge \tau \rangle, \quad \forall\, X \in TM,\ \omega \in \Lambda^k M,\ \tau \in \Lambda^{k-1}M. \quad (14.3)$$

We define the Hodge $*$-operator $* : \Lambda^k M \to \Lambda^{n-k} M$ by

$$\omega \wedge *\tau := \langle \omega, \tau \rangle dv, \quad \forall\, \omega, \tau \in \Lambda^k M.$$

It is well known and easy to check on the local basis above that the following relations are satisfied:

$$*1 = dv, \quad *dv = 1, \quad (14.4)$$

$$\langle *\omega, *\tau \rangle = \langle \omega, \tau \rangle, \quad (14.5)$$

$$*^2 = (-1)^{k(n-k)} \quad \text{on } \Lambda^k M. \tag{14.6}$$

The exterior derivative $d : \Omega^k M \to \Omega^{k+1} M$, which by (6.5) satisfies

$$d = \sum_{i=1}^{n} e_i \wedge \nabla_{e_i},$$

has a formal adjoint $\delta : \Omega^{k+1} M \to \Omega^k M$ given by

$$\delta = -(-1)^{nk} * d* = - \sum_{i=1}^{n} e_i \lrcorner \nabla_{e_i}. \tag{14.7}$$

To see this, let $\alpha \in \Omega^k$ and $\beta \in \Omega^{k+1}$ be smooth forms. Then we have

$$\begin{aligned}
\langle d\alpha, \beta \rangle dv &= d\alpha \wedge *\beta = d(\alpha \wedge *\beta) - (-1)^k \alpha \wedge d * \beta \\
&= d(\alpha \wedge *\beta) - (-1)^{k+k(n-k)} \alpha \wedge * * d * \beta \\
&= d(\alpha \wedge *\beta) - (-1)^{nk} \langle \alpha, *d*\beta \rangle dv,
\end{aligned}$$

so Lemma 14.3 shows that $d^* = (-1)^{nk+1} * d*$ on $k + 1$-forms. The formal adjoint $\delta := d^*$ of d is called the *codifferential* and a form ω which satisfies $\delta \omega = 0$ is called *coclosed*.

Using the Hodge $*$-operator we get the following useful reformulation of Lemma 14.3: if there exists a section $\tau \in \Gamma(E^* \otimes F^* \otimes \Lambda^1 M)$ such that

$$\langle P\alpha, \beta \rangle_F - \langle \alpha, Q\beta \rangle_E = \delta(\tau(\alpha, \beta)), \quad \forall \, \alpha \in \Gamma(E), \, \beta \in \Gamma(F), \tag{14.8}$$

then Q is the formal adjoint of P. Indeed, $\delta(\tau(\alpha, \beta)) dv = -d(*(\tau(\alpha, \beta)))$ is an exact n-form.

The *Laplace operator* $\Delta : \Omega^k M \to \Omega^k M$ is defined by

$$\Delta := d\delta + \delta d,$$

and is clearly formally self-adjoint. An exterior form $\omega \in \Omega^p M$ is called *harmonic* if $\Delta \omega = 0$.

14.3. The Laplace operator on Kähler manifolds

After these preliminaries, let now (M^{2m}, h, J) be an almost Hermitian manifold with fundamental form Ω. We define the following (real) algebraic operators acting on differential forms:

$$L : \Lambda^k M \to \Lambda^{k+2} M, \qquad L(\omega) := \Omega \wedge \omega = \frac{1}{2} \sum_{i=1}^{2m} e_i \wedge Je_i \wedge \omega,$$

with adjoint Λ satisfying

$$\Lambda : \Lambda^{k+2} M \to \Lambda^k M, \qquad \Lambda(\omega) := \frac{1}{2} \sum_{i=1}^{2m} Je_i \lrcorner e_i \lrcorner \omega.$$

These natural operators can be extended to complex-valued forms by $\mathbb{C}$-linearity. If P and Q are differential or algebraic operators acting on sections of the same vector bundle, we denote by a bracket the operator $[P, Q] := P \circ Q - Q \circ P$.

LEMMA 14.4. *The following relations hold:*

(1) *The Hodge $*$-operator maps (p, q)-forms to $(m - q, m - p)$-forms.*
(2) $[X \lrcorner, \Lambda] = 0$ *and* $[X \lrcorner, L] = JX \wedge .$

The proof is straightforward.

Let us now assume that M is Kähler. We define the twisted differential $d^c : \Omega^k M \to \Omega^{k+1} M$ by

$$d^c(\omega) := \sum_{i=1}^{2m} Je_i \wedge \nabla_{e_i}\omega$$

whose formal adjoint is $\delta^c : \Omega^{k+1} M \to \Omega^k M$

$$\delta^c := - * d^c * = - \sum_{i=1}^{2m} Je_i \lrcorner \nabla_{e_i}.$$

LEMMA 14.5. *On a Kähler manifold, the following relations, called* Kähler identities *hold:*

$$[L, \delta] = d^c, \qquad [L, d] = 0 \tag{14.9}$$

and

$$[\Lambda, d] = -\delta^c, \qquad [\Lambda, \delta] = 0. \tag{14.10}$$

PROOF. Using Lemma 14.4(2) and the fact that J and Ω are parallel we get

$$[L, \delta] = - \sum_{i=1}^{2m} [L, e_i \lrcorner \nabla_{e_i}] = - \sum_{i=1}^{2m} [L, e_i \lrcorner]\nabla_{e_i} = \sum_{i=1}^{2m} Je_i \wedge \nabla_{e_i} = d^c.$$

The second relation in (14.9) follows from the fact that the Kähler form is closed. The two relations in (14.10) are direct consequences of (14.9) using the Hodge $*$-operator. □

Corresponding to the decomposition $d = \partial + \bar{\partial}$ we have the decomposition $\delta = \partial^* + \bar{\partial}^*$, where

$$\partial^* : \Omega^{p,q} M \to \Omega^{p-1,q} M, \qquad \partial^* := - * \bar{\partial} *$$

and

$$\bar{\partial}^* : \Omega^{p,q} M \to \Omega^{p,q-1} M, \qquad \bar{\partial}^* := - * \partial * .$$

Notice that ∂^* and $\bar{\partial}^*$ are formal adjoints of ∂ and $\bar{\partial}$ with respect to the Hermitian product H on complex forms given by

$$H(\omega, \tau) := \langle \omega, \bar{\tau} \rangle. \tag{14.11}$$

We define the Laplace operators

$$\Delta^\partial := \partial\partial^* + \partial^*\partial \qquad \text{and} \qquad \Delta^{\bar\partial} := \bar\partial\bar\partial^* + \bar\partial^*\bar\partial.$$

One of the most important features of Kähler metrics is that these new Laplace operators are essentially the same as the usual one:

THEOREM 14.6. *On any Kähler manifold one has* $\Delta = 2\Delta^\partial = 2\Delta^{\bar\partial}$.

PROOF. The usual identification of TM and T^*M via the metric, extended to $T^{\mathbb{C}}M$ by $\mathbb{C}$-linearity, maps $(1,0)$-vectors to $(0,1)$-forms and vice versa. From the fact that $\Lambda^{p,q}M$ are parallel subbundles with respect to the covariant derivative ∇, we easily get

$$\partial = \sum_j \frac{1}{2}(e_j + iJe_j) \wedge \nabla_{e_j} \qquad \text{and} \qquad \bar\partial = \sum_j \frac{1}{2}(e_j - iJe_j) \wedge \nabla_{e_j}.$$

Subtracting these two relations yields

$$d^c = i(\bar\partial - \partial), \tag{14.12}$$

and applying the Hodge $*$-operator,

$$\delta^c = i(\partial^* - \bar\partial^*). \tag{14.13}$$

Applying (14.9) to a (p,q)-form and projecting onto $\Lambda^{p\pm1,q}M$ and $\Lambda^{p,q\pm1}M$ give

$$[L, \partial^*] = i\bar\partial, \qquad [L, \bar\partial^*] = -i\partial, \qquad [L, \partial] = 0, \qquad [L, \bar\partial] = 0, \tag{14.14}$$

and similarly from (14.10) we obtain

$$[\Lambda, \partial] = i\bar\partial^*, \qquad [\Lambda, \bar\partial] = -i\partial^*, \qquad [\Lambda, \partial^*] = 0, \qquad [\Lambda, \bar\partial^*] = 0. \tag{14.15}$$

Now, the relation $\bar\partial^2 = 0$ together with (14.15) enables us to write

$$-i(\bar\partial\partial^* + \partial^*\bar\partial) = \bar\partial[\Lambda, \bar\partial] + [\Lambda, \bar\partial]\bar\partial = \bar\partial\Lambda\bar\partial - \bar\partial\Lambda\bar\partial = 0$$

and similarly

$$\partial\bar\partial^* + \bar\partial^*\partial = 0.$$

Thus,

$$\begin{aligned} \Delta &= (\partial + \bar\partial)(\partial^* + \bar\partial^*) + (\partial^* + \bar\partial^*)(\partial + \bar\partial) \\ &= (\partial\partial^* + \partial^*\partial) + (\bar\partial\bar\partial^* + \bar\partial^*\bar\partial) + (\bar\partial\partial^* + \partial^*\bar\partial) + (\partial\bar\partial^* + \bar\partial^*\partial) \\ &= \Delta^\partial + \Delta^{\bar\partial}. \end{aligned}$$

It remains to show the equality $\Delta^\partial = \Delta^{\bar\partial}$, which is a consequence of (14.15):

$$\begin{aligned} -i\Delta^\partial &= -i(\partial\partial^* + \partial^*\partial) = \partial[\Lambda, \bar\partial] + [\Lambda, \bar\partial]\partial = \partial\Lambda\bar\partial - \partial\bar\partial\Lambda + \Lambda\bar\partial\partial - \bar\partial\Lambda\partial \\ &= \partial\Lambda\bar\partial + \bar\partial\partial\Lambda - \Lambda\partial\bar\partial - \bar\partial\Lambda\partial = [\partial, \Lambda]\bar\partial + \bar\partial[\partial, \Lambda] \\ &= -i\bar\partial^*\bar\partial - i\bar\partial\bar\partial^* = -i\Delta^{\bar\partial}. \end{aligned}$$

$\square$

14.4. Exercises

(1) Consider the extension of J as derivation

$$J : \Lambda^k M \to \Lambda^k M, \qquad J(\omega) := \sum_{i=1}^{2m} Je_i \wedge (e_i \lrcorner \omega).$$

Show that the following relations hold:
- J is skew-Hermitian.
- $J(\alpha \wedge \beta) = J(\alpha) \wedge \beta + \alpha \wedge J(\beta)$ for all forms $\alpha \in \Omega^p M$ and $\beta \in \Omega^k M$.
- The restriction of J to $\Lambda^{p,q} M$ is equal to the scalar multiplication by $i(q - p)$.
- $[J, \Lambda] = 0$ and $[J, L] = 0$.

(2) Let ω be a k-form on an n-dimensional Riemannian manifold M. Prove that

$$\sum_{i=1}^{n} e_i \wedge (e_i \lrcorner \omega) = k\omega \qquad \text{and} \qquad \sum_{i=1}^{n} e_i \lrcorner (e_i \wedge \omega) = (n - k)\omega.$$

(3) Show that $0 = dd^c + d^c d = d\delta^c + \delta^c d = \delta\delta^c + \delta^c\delta = \delta d^c + d^c\delta$ on every Kähler manifold.

(4) Prove that $[J, d] = d^c$ and $[J, d^c] = -d$ on Kähler manifolds.

(5) Show that the Laplace operator commutes with L, Λ and J on Kähler manifolds.

CHAPTER 15

Hodge and Dolbeault theories

15.1. Hodge theory

In this section we assume that (M^n, g) is a compact oriented Riemannian manifold. From now on we denote by $\Omega_{\mathbb{C}}^k M := \Gamma(\Lambda^k M \otimes \mathbb{C})$ the space of smooth complex-valued k-forms and by $\mathcal{Z}_{\mathbb{C}}^k M$ the space of *closed* complex k-forms on M. Since the exterior derivative satisfies $d^2 = 0$, one clearly has $d\Omega_{\mathbb{C}}^{k-1} M \subset \mathcal{Z}_{\mathbb{C}}^k M$. We define the *de Rham cohomology groups* by

$$H_{DR}^k(M, \mathbb{C}) := \frac{\mathcal{Z}_{\mathbb{C}}^k M}{d\Omega_{\mathbb{C}}^{k-1} M}.$$

The *de Rham isomorphism theorem* says that the kth singular cohomology group of M with complex coefficients is naturally isomorphic to the kth de Rham cohomology group:

$$H^k(M, \mathbb{C}) \simeq H_{DR}^k(M, \mathbb{C}).$$

We now denote by $\mathcal{H}^k(M, \mathbb{C})$ the space of complex *harmonic k-forms* on M, i.e. forms in the kernel of the Laplace operator:

$$\mathcal{H}^k(M, \mathbb{C}) := \{\omega \in \Omega_{\mathbb{C}}^k M \mid \Delta\omega = 0\}.$$

LEMMA 15.1. *A form is harmonic if and only if it is closed and δ-closed.*

PROOF. One direction is clear. Suppose conversely that ω is harmonic. Since M is compact and d and δ are formally adjoint operators with respect to the Hermitian product H defined in (14.11), we get

$$0 = \int_M H(\Delta\omega, \omega)dv = \int_M H(d\delta\omega + \delta d\omega, \omega)dv = \int_M |\delta\omega|^2 + |d\omega|^2 dv,$$

showing that $d\omega = 0$ and $\delta\omega = 0$. $\qquad\square$

THEOREM 15.2. (Hodge decomposition theorem) *The space of k-forms decomposes as a direct sum*

$$\Omega_{\mathbb{C}}^k M = \mathcal{H}^k(M, \mathbb{C}) \oplus \delta\Omega_{\mathbb{C}}^{k+1} M \oplus d\Omega_{\mathbb{C}}^{k-1} M.$$

PROOF. Using Lemma 15.1 it can be checked immediately that the three spaces above are orthogonal with respect to the global Hermitian product on

$\Omega_{\mathbb{C}}^k M$ given by

$$(\omega, \tau) := \int_M H(\omega, \tau) dv.$$

The hard part of the theorem is to show that the direct sum of these three summands is the whole space $\Omega_{\mathbb{C}}^k M$. A proof can be found in [3], pp. 84–100. □

The Hodge decomposition theorem shows that every complex k-form ω on M can be uniquely written as

$$\omega = d\omega' + \delta\omega'' + \omega^H,$$

where $\omega' \in \Omega_{\mathbb{C}}^{k-1} M$, $\omega'' \in \Omega_{\mathbb{C}}^{k+1} M$ and $\omega^H \in \mathcal{H}^k(M, \mathbb{C})$. The above expression is called the *Hodge decomposition* of ω. If ω is closed, we can write

$$0 = (d\omega, \omega'') = (d\delta\omega'', \omega'') = \int_M |\delta\omega''|^2 dv,$$

showing that the second term in the Hodge decomposition of ω vanishes.

PROPOSITION 15.3. (Hodge isomorphism) *The natural map*

$$f : \mathcal{H}^k(M, \mathbb{C}) \to H_{DR}^k(M, \mathbb{C})$$

given by $\omega \mapsto [\omega]$ *is an isomorphism.*

PROOF. First, f is well-defined because every harmonic form is closed (Lemma 15.1). The kernel of f is zero since the space of harmonic forms is H-orthogonal to the space of exact forms, so in particular their intersection is reduced to $\{0\}$. Finally, for every de Rham cohomology class c, let ω be a closed form such that $[\omega] = c$. We have seen that the Hodge decomposition of ω is $\omega = d\omega' + \omega^H$, showing that

$$f(\omega^H) = [\omega^H] = [d\omega' + \omega^H] = [\omega] = c,$$

hence f is onto. □

The complex dimension $b_k(M) := \dim_{\mathbb{C}}(H_{DR}^k(M, \mathbb{C}))$ is called the kth Betti number of M and is a topological invariant by de Rham's theorem.

PROPOSITION 15.4. (Poincaré duality) *The vector spaces* $\mathcal{H}^k(M, \mathbb{C})$ *and* $\mathcal{H}^{n-k}(M, \mathbb{C})$ *are isomorphic. In particular* $b_k(M) = b_{n-k}(M)$ *for every compact n-dimensional manifold M.*

PROOF. The isomorphism is simply given by the Hodge $*$-operator which maps harmonic k-forms to harmonic $(n-k)$-forms. □

We close this section with the following interesting application of Theorem 15.2.

PROPOSITION 15.5. *Every Killing vector field on a compact Kähler manifold (M, g, J) is real holomorphic.*

PROOF. Let X be a Killing vector field, that is, by Proposition 6.9, $\mathcal{L}_X g = 0$. We compute the Lie derivative of the Kähler 2-form with respect to X using the Cartan formula:

$$\mathcal{L}_X \Omega = d(X \lrcorner \Omega) + X \lrcorner d\Omega = d(X \lrcorner \Omega),$$

so $\mathcal{L}_X \Omega$ is exact. On the other hand, since the flow of X is isometric, it commutes with the Hodge $*$-operator, thus $\mathcal{L}_X \circ * = * \circ \mathcal{L}_X$. As we clearly have $d \circ \mathcal{L}_X = \mathcal{L}_X \circ d$ too, we see that $\mathcal{L}_X \circ \delta = \delta \circ \mathcal{L}_X$, whence

$$d(\mathcal{L}_X \Omega) = \mathcal{L}_X (d\Omega) = 0$$

and

$$\delta(\mathcal{L}_X \Omega) = \mathcal{L}_X (\delta\Omega) = 0,$$

because Ω, being parallel, is coclosed. Thus $\mathcal{L}_X \Omega$ is harmonic and exact, so it has to vanish by Theorem 15.2. This shows that the flow of X preserves the metric and the Kähler 2-form, it thus preserves the complex structure J, hence X is real holomorphic. □

REMARK. The above result does not hold without the compactness assumption, see Exercise (5) below.

15.2. Dolbeault theory

Let (M^{2m}, h, J) be a compact Hermitian manifold. We consider the Dolbeault operator $\bar{\partial}$ acting on the spaces of (p, q)-forms $\Omega^{p,q} M := \Gamma(\Lambda^{p,q} M) \subset \Omega_{\mathbb{C}}^{p+q} M$. Let $\mathcal{Z}^{p,q} M$ denote the space of $\bar{\partial}$-closed (p, q)-forms. Since $\bar{\partial}^2 = 0$, we see that $\bar{\partial}\Omega^{p,q-1} M \subset \mathcal{Z}^{p,q} M$. We define the Dolbeault cohomology groups

$$H^{p,q} M := \frac{\mathcal{Z}^{p,q} M}{\bar{\partial}\Omega^{p,q-1} M}.$$

In contrast to de Rham cohomology, the Dolbeault cohomology is no longer a topological invariant of the manifold, since it strongly depends on the complex structure J.

We define the space $\mathcal{H}^{p,q} M$ of $\bar{\partial}$-*harmonic* (p, q)-*forms* on M by

$$\mathcal{H}^{p,q} M := \{\omega \in \Omega^{p,q} M \mid \Delta^{\bar{\partial}} \omega = 0\}.$$

As before we have:

LEMMA 15.6. *A form $\omega \in \Omega^{p,q} M$ is $\bar{\partial}$-harmonic if and only if $\bar{\partial}\omega = 0$ and $\bar{\partial}^* \omega = 0$.*

The proof is very similar to that of Lemma 15.1 and is left as an exercise.

THEOREM 15.7. (Dolbeault decomposition theorem) *The space of (p,q)-forms decomposes as a direct sum*

$$\Omega^{p,q} M = \mathcal{H}^{p,q} M \oplus \bar{\partial}^* \Omega^{p,q+1} M \oplus \bar{\partial} \Omega^{p,q-1} M.$$

PROOF. Lemma 15.6 shows that the three spaces above are orthogonal with respect to the global Hermitian product

$$(\cdot, \cdot) := \int_M H(\cdot, \cdot) dv$$

on $\Omega^{p,q} M$. A proof for the hard part, which consists in showing that the direct sum of the three summands is the whole space $\Omega^{p,q} M$, can be found in [3], pp. 84–100. □

Every (p,q)-form ω on M can thus be uniquely written as

$$\omega = \bar{\partial} \omega' + \bar{\partial}^* \omega'' + \omega^H,$$

where $\omega' \in \Omega^{p,q-1} M$, $\omega'' \in \Omega^{p,q+1} M$ and $\omega^H \in \mathcal{H}^{p,q} M$. This is called the *Dolbeault decomposition* of ω. As before, the second summand in the Dolbeault decomposition of ω vanishes if and only if $\bar{\partial}\omega = 0$. Specializing for $q = 0$ yields:

PROPOSITION 15.8. *A $(p,0)$-form on a compact Hermitian manifold is holomorphic if and only if it is $\bar{\partial}$-harmonic.*

COROLLARY 15.9. (Dolbeault isomorphism) *The map $\mathcal{H}^{p,q} M \to H^{p,q} M$ given by $\omega \mapsto [\omega]$ is an isomorphism.*

The proof is completely similar to the proof of the Hodge isomorphism.

We denote by $h^{p,q}$ the complex dimension of $H^{p,q} M$. These are the *Hodge numbers* associated to the complex structure J of M.

PROPOSITION 15.10. (Serre duality) *The spaces $\mathcal{H}^{p,q} M$ and $\mathcal{H}^{m-p,m-q} M$ are isomorphic. In particular $h^{p,q} = h^{m-p,m-q}$.*

PROOF. Consider the composition of the Hodge $*$-operator with the complex conjugation

$$\bar{*} : \Omega^{p,q} M \to \Omega^{m-p,m-q} M, \qquad \bar{*}\omega := *\bar{\omega}.$$

We have

$$
\begin{aligned}
\bar{*}\Delta^{\bar{\partial}}(\omega) &= \overline{*(\bar{\partial}\bar{\partial}^* + \bar{\partial}^*\bar{\partial})\omega} = *(\partial\partial^* + \partial^*\partial)\bar{\omega} \\
&= -*(\partial * \bar{\partial} * + * \bar{\partial} * \partial)\bar{\omega} = \bar{\partial}^*\bar{\partial}(\bar{*}\omega) - *^2\bar{\partial} * \partial\bar{\omega} \\
&= \bar{\partial}^*\bar{\partial}(\bar{*}\omega) - \bar{\partial} * \partial *^2 \bar{\omega} = \bar{\partial}^*\bar{\partial}(\bar{*}\omega) + \bar{\partial}\bar{\partial}^*(\bar{*}\omega) = \Delta^{\bar{\partial}}(\bar{*}\omega).
\end{aligned}
$$

This clearly shows that $\bar{*}$ is a ($\mathbb{C}$-anti-linear) isomorphism from $\mathcal{H}^{p,q} M$ to $\mathcal{H}^{m-p,m-q} M$. □

If M is Kähler, much more can be said about Hodge and Betti numbers, due to Theorem 14.6. Firstly, the fact that $\Delta = 2\Delta^{\bar{\partial}}$ shows that $\mathcal{H}^{p,q}M \subset \mathcal{H}^{p+q}M$. Secondly, since $\Delta^{\bar{\partial}}$ leaves the spaces $\Omega^{p,q}M$ invariant, we deduce that Δ has the same property, thus proving that the components of a harmonic form in its type decomposition are all harmonic. This gives the following direct sum decomposition:

$$\mathcal{H}^k M = \bigoplus_{p+q=k} \mathcal{H}^{p,q} M.$$

Moreover, as $\Delta^{\bar{\partial}} = \frac{1}{2}\Delta$ is a real operator on Kähler manifolds, it commutes with the complex conjugation (in the general case we only have that $\overline{\Delta^{\bar{\partial}}\alpha} = \Delta^{\partial}\bar{\alpha}$) so the complex conjugation defines an isomorphism between the spaces $\mathcal{H}^{p,q}M$ and $\mathcal{H}^{q,p}M$. Consider now the Kähler form $\Omega \in \Omega^{1,1}M$. Since Ω^m is a non-zero multiple of the volume form, we deduce that all exterior powers $\Omega^p \in \Omega^{p,p}M$ are non-zero. Moreover, they are all harmonic since every exterior form parallel with respect to the Levi–Civita covariant derivative is automatically harmonic. Summarizing, we have proved the

PROPOSITION 15.11. *In addition to Poincaré and Serre dualities, the following relations hold between Betti and Hodge numbers on compact Kähler manifolds:*

$$b_k = \sum_{p+q=k} h^{p,q}, \qquad h^{p,q} = h^{q,p}, \qquad h^{p,p} \geq 1, \quad \forall\, 0 \leq p \leq m. \qquad (15.1)$$

In particular (15.1) *shows that all Betti numbers of odd order are even and all Betti numbers of even order are non-zero.*

15.3. Exercises

(1) Prove that the complex manifold $S^1 \times S^{2k+1}$ carries no Kähler metric for $k \geq 1$. *Hint:* Use the formula for the Betti numbers of a product:

$$b_k(M \times N) = \sum_{p+q=k} b_p(M)b_q(N).$$

(2) *The global $i\partial\bar{\partial}$-lemma.* Let φ be an exact real $(1,1)$-form on a compact Kähler manifold M. Prove that φ is $i\partial\bar{\partial}$-exact, in the sense that there exists a real function u such that $\varphi = i\partial\bar{\partial}u$.

(3) Show that there exists no global Kähler potential on a compact Kähler manifold.

(4) Let (M, h, J) be a compact Kähler manifold whose second Betti number is equal to 1. Show that if the scalar curvature of M is constant, then the metric h is Einstein. *Hint:* Use formula (12.7).

(5) Consider the Kähler manifold $M = (\mathbb{R}^{2m}, g, J)$, where g and J are the standard Euclidean and complex structure on $\mathbb{R}^{2m}$. For every matrix $A \in \mathcal{M}_{2m}(\mathbb{R})$ we define a vector field ξ^A on $\mathbb{R}^{2m}$ by

$$\xi^A_x := Ax, \qquad \forall\, x \in \mathbb{R}^{2m}.$$

Show that the local flow of ξ^A is given by $\varphi_t = \exp(tA)$. Deduce that ξ^A is Killing if and only if A is skew-symmetric and ξ^A is real holomorphic if and only if A commutes with j_m (defined in (7.1)).

Part 3

Topics on compact Kähler manifolds

CHAPTER 16

Chern classes

16.1. Chern–Weil theory

The comprehensive theory of Chern classes can be found in [11], Ch. 12. We will outline here the definition and properties of the first Chern class, which is the only one needed in the sequel. The following proposition can be taken as a definition:

PROPOSITION 16.1. *To every complex vector bundle E over a smooth manifold M one can associate a cohomology class $c_1(E) \in H^2(M, \mathbb{Z})$ called the first Chern class of E satisfying the following axioms:*

- (Naturality) *For every smooth map $f : M \to N$ and complex vector bundle E over N, one has $f^*(c_1(E)) = c_1(f^*E)$, where the left term denotes the pull-back in cohomology and f^*E is the pull-back bundle defined by $f^*E_x = E_{f(x)}, \forall\, x \in M$.*
- (Whitney sum formula) *For every bundles E, F over M one has $c_1(E \oplus F) = c_1(E) + c_1(F)$, where $E \oplus F$ is the Whitney sum defined as the pull-back of the bundle $E \times F \to M \times M$ by the diagonal inclusion of M in $M \times M$.*
- (Normalization) *The first Chern class of the tautological bundle of $\mathbb{CP}^1$ is equal to -1 in $H^2(\mathbb{CP}^1, \mathbb{Z}) \simeq \mathbb{Z}$, which means that the integral over $\mathbb{CP}^1$ of any representative of this class equals -1.*

Let $E \to M$ be a complex vector bundle. We will now explain the *Chern–Weil theory*, which allows one to express the images in real cohomology of the Chern classes of E using the curvature of an arbitrary connection ∇ on E. Recall the formula (10.2) for the curvature of ∇:

$$R^\nabla(\sigma_i) = \sum_{j=1}^k R_{ij}^\nabla \sigma_j = \sum_{j=1}^k \left(d\omega_{ij} - \sum_{l=1}^k \omega_{il} \wedge \omega_{lj}\right)\sigma_j, \qquad (16.1)$$

where $\{\sigma_1, \ldots, \sigma_k\}$ are local sections of E which form a basis of each fibre over some open set U and the connection forms $\omega_{ij} \in \Lambda^1(U)$ (relative to the choice of this basis) are defined by

$$\nabla \sigma_i = \sum_{j=1}^k \omega_{ij} \otimes \sigma_j.$$

Notice that although the coefficients R^∇_{ij} of R^∇ depend on the local basis of sections (σ_i), its trace is a well-defined (complex-valued) 2-form on M, independent of the chosen basis, and can be expressed as $\text{Tr}(R^\nabla) = \sum R^\nabla_{ii}$ in the local basis (σ_i). To compute it explicitly, we use the following summation trick:

$$\sum_{i,l=1}^{k} \omega_{il} \wedge \omega_{li} = \sum_{l,i=1}^{k} \omega_{li} \wedge \omega_{il} = -\sum_{i,l=1}^{k} \omega_{il} \wedge \omega_{li},$$

where the first equality is given by interchanging the summation subscripts and the second by the fact that the wedge product is skew-symmetric on 1-forms. From (16.1) we thus get

$$\text{Tr}(R^\nabla) = d\left(\sum \omega_{ii}\right), \tag{16.2}$$

where of course the trace of the connection form $\omega = (\omega_{ij})$ *does depend* on the local basis (σ_i). This shows that $\text{Tr}(R^\nabla)$ is closed, being locally exact.

If ∇ and $\tilde\nabla$ are connections on E, the Leibniz rule shows that their difference $A := \tilde\nabla - \nabla$ is a zero-order operator, more precisely a smooth section of $\Lambda^1(M) \otimes \text{End}(E)$. Thus $\text{Tr}(A)$ is a well-defined 1-form on M and (16.2) readily implies

$$\text{Tr}(R^{\tilde\nabla}) = \text{Tr}(R^\nabla) + d(\text{Tr}(A)). \tag{16.3}$$

We thus have proved the following:

LEMMA 16.2. *The cohomology class* $[\text{Tr}(R^\nabla)] \in H^2(M, \mathbb{C})$ *of the closed 2-form* $\text{Tr}(R^\nabla)$ *does not depend on* ∇.

It is actually easy to see that $[\text{Tr}(R^\nabla)]$ is a purely imaginary class, in the sense that it has a representative which is a purely imaginary 2-form. Indeed, let us choose an arbitrary Hermitian structure h on E and take ∇ such that h is ∇-parallel. If we start with a local basis $\{\sigma_i\}$ adapted to h, then we have

$$\begin{aligned}
0 &= \nabla(\delta_{ij}) = \nabla(h(\sigma_i, \sigma_j)) = h(\nabla\sigma_i, \sigma_j) + h(\sigma_i, \nabla\sigma_j) \\
&= \omega_{ij} + \overline{\omega_{ji}}.
\end{aligned}$$

From (16.1) we get

$$\begin{aligned}
\overline{R^\nabla_{ij}} &= d\overline{\omega_{ij}} - \sum_{l=1}^{k} \overline{\omega_{il}} \wedge \overline{\omega_{lj}} = -d\omega_{ji} - \sum_{l=1}^{k} \omega_{li} \wedge \omega_{jl} \\
&= -d\omega_{ji} + \sum_{l=1}^{k} \omega_{jl} \wedge \omega_{li} = -R^\nabla_{ji},
\end{aligned}$$

showing that the trace of R^∇ is a purely imaginary 2-form.

THEOREM 16.3. *Let ∇ be a connection on a complex bundle E over M. The real cohomology class*

$$c_1(\nabla) := \left[\frac{i}{2\pi}\mathrm{Tr}(R^\nabla)\right]$$

is equal to the image of $c_1(E)$ in $H^2(M, \mathbb{R})$.

PROOF. We have to check that $c_1(\nabla)$ satisfies the three conditions in Proposition 16.1. In order to prove the naturality, recall that if $f : M \to N$ is smooth and $\pi : E \to N$ is a rank k vector bundle, then

$$f^*(E) := \{(x, v) \mid x \in M,\ v \in E,\ f(x) = \pi(v)\}.$$

If $\{\sigma_i\}$ is a local basis of sections of E, then

$$f^*\sigma_i : M \to f^*(E), \qquad x \mapsto (x, \sigma_i(f(x)))$$

is a basis of local sections of f^*E. The formula

$$f^*\nabla(f^*\sigma) := f^*(\nabla\sigma)$$

defines a connection on f^*E (see Section 5.4), and with respect to the local basis $(f^*\sigma_i)$, we obviously have

$$R_{ij}^{f^*\nabla} = f^*(R_{ij}^\nabla),$$

whence $c_1(f^*\nabla) = f^*(c_1(\nabla))$.

The Whitney sum formula is also easy to check. If E and F are complex bundles over M with connections ∇ and $\tilde{\nabla}$ then one can define a connection $\nabla \oplus \tilde{\nabla}$ on $E \oplus F$ by

$$(\nabla \oplus \tilde{\nabla})_X(\sigma \oplus \tilde{\sigma}) := \nabla_X\sigma \oplus \tilde{\nabla}_X\tilde{\sigma}.$$

If $\{\sigma_i\}$, $\{\tilde{\sigma}_j\}$ are local bases of sections of E and F then $\{\sigma_i \oplus 0, 0 \oplus \tilde{\sigma}_j\}$ is a local basis for $E \oplus F$ and the curvature of $\nabla \oplus \tilde{\nabla}$ in this basis is a block matrix having R^∇ and $R^{\tilde{\nabla}}$ on the principal diagonal. Its trace is thus the sum of the traces of R^∇ and $R^{\tilde{\nabla}}$.

We finally check the normalization property. Let $L \to \mathbb{CP}^1$ be the tautological bundle. For any section $\sigma : \mathbb{CP}^1 \to L$ of L we denote by $\sigma_0 : U_0 \to \mathbb{C}$ and $\sigma_1 : U_1 \to \mathbb{C}$ the expressions of σ in the standard local trivializations of L, given by $\psi_i : \pi^{-1}U_i \to U_i \times \mathbb{C}$, $\psi_i(w) = (\pi(w), w_i)$.

The Hermitian product on $\mathbb{C}^2$ induces a Hermitian structure h on L. Let ∇ be the Chern connection on L associated to h. We choose a local holomorphic section σ and denote its square norm by u. If ω is the connection form of ∇ with respect to the section σ (i.e. $\nabla\sigma = \omega \otimes \sigma$), then for all $X \in T\mathbb{CP}^1$ we can write:

$$\partial_X(u) = \partial_X(h(\sigma, \sigma)) = h(\nabla_X\sigma, \sigma) + h(\sigma, \nabla_X\sigma) = \omega(X)u + \bar{\omega}(X)u.$$

This just means $\omega + \bar{\omega} = d \log u$. On the other hand, since σ is holomorphic and $\nabla^{0,1} = \bar{\partial}$, we see that ω is a $(1,0)$-form. Thus $\omega = \partial \log u$. From (16.2) we get

$$R^{\nabla} = d\omega = d\partial \log u = \bar{\partial} \partial \log u. \tag{16.4}$$

The normalization axiom is thus equivalent to the following relation:

$$\frac{i}{2\pi} \int_{\mathbb{CP}^1} \bar{\partial} \partial \log u = -1.$$

It is sufficient to compute this integral over $U_0 := \mathbb{CP}^1 \backslash \{[0 : 1]\}$. We denote by $z := \phi_0 = z_1/z_0$ the holomorphic coordinate on U_0. Consider the holomorphic section σ over U_0 which satisfies $\sigma_0 \equiv 1$. From the definition of σ_0 (as the image of σ through the trivialization ψ_0 of L), we deduce that $\sigma(z)$ is the unique vector lying on the complex line generated by (z_0, z_1) in $\mathbb{C}^2$, whose first coordinate is 1, i.e. $\sigma(z) = (1, z)$. This shows that $u = |(1, z)|^2 = 1 + |z|^2$. In polar coordinates $z = r \cos \theta + ir \sin \theta$ one can readily compute for every function $f(r, \theta)$:

$$\bar{\partial} \partial f = \frac{i}{2} \left(r \frac{\partial^2 f}{\partial r^2} + \frac{1}{r} \frac{\partial^2 f}{\partial \theta^2} + \frac{\partial f}{\partial r} \right) dr \wedge d\theta.$$

Applying this formula to $f := \log(1 + r^2)$ we finally get

$$\begin{aligned}
\frac{i}{2\pi} \int_{\mathbb{CP}^1} \bar{\partial} \partial \log u &= \frac{i}{2\pi} \int_{[0,\infty) \times [0,2\pi]} \frac{i}{2} \left(r \frac{\partial^2 f}{\partial r^2} + \frac{\partial f}{\partial r} \right) dr \wedge d\theta \\
&= -\frac{1}{2} \int_0^\infty d \left(r \frac{\partial f}{\partial r} \right) = \frac{1}{2} \lim_{r \to \infty} r \frac{\partial f}{\partial r} \\
&= -\lim_{r \to \infty} \frac{r}{2} \frac{2r}{1 + r^2} = -1.
\end{aligned}$$

$\square$

If M is an almost complex manifold, we define the first Chern class of M — denoted by $c_1(M)$ — to be the first Chern class of the tangent bundle TM, viewed as complex vector bundle:

$$c_1(M) := c_1(TM).$$

In Section 17.2 below, we will see that a representative of the first Chern class of a Kähler manifold is $(1/2\pi)\rho$, where ρ denotes the Ricci form.

16.2. Properties of the first Chern class

Let M be a smooth manifold and let E, F be two complex vector bundles over M.

Proposition 16.4. (i) $c_1(E) = c_1(\Lambda^k E)$, where k denotes the rank of E.

(ii) $c_1(E \otimes F) = rk(F)c_1(E) + rk(E)c_1(F)$.

(iii) $c_1(E^*) = -c_1(E)$, where E^* denotes the dual of E.

Proof. (i) Consider any connection ∇ on E, inducing a connection $\tilde{\nabla}$ on $\Lambda^k E$. If $\sigma_1, \ldots, \sigma_k$ denotes a local basis of sections of E, then $\sigma := \sigma_1 \wedge \cdots \wedge \sigma_k$ is a local non-vanishing section of $\Lambda^k E$. Let $\omega := (\omega_{ij})$ and $\tilde{\omega}$ be the connection forms of ∇ and $\tilde{\nabla}$ relative to these local bases:

$$\nabla \sigma_i = \omega_{ij} \otimes \sigma_j \qquad \text{and} \qquad \tilde{\nabla}\sigma = \tilde{\omega} \otimes \sigma.$$

We compute

$$
\begin{aligned}
\tilde{\nabla}\sigma &= \tilde{\nabla}(\sigma_1 \wedge \cdots \wedge \sigma_k) \\
&= \sum_{i=1}^{k} \sigma_1 \wedge \cdots \wedge \sigma_{i-1} \wedge \Big(\sum_{j=1}^{k} \omega_{ij} \otimes \sigma_j \Big) \wedge \sigma_{i+1} \wedge \cdots \wedge \sigma_k \\
&= \sum_{i=j} \omega_{ij} \otimes \sigma,
\end{aligned}
$$

which proves that $\tilde{\omega} = \text{Tr}(\omega)$. From (16.1) we then get

$$\text{Tr}(R^{\tilde{\nabla}}) = R^{\tilde{\nabla}} = d\tilde{\omega} - \tilde{\omega} \wedge \tilde{\omega} = d\tilde{\omega}$$

and

$$\text{Tr}(R^{\nabla}) = \sum_{i=j}(d\omega_{ij} - \omega_{ik} \wedge \omega_{kj}) = \sum_{i=j} d\omega_{ij} = d\text{Tr}(\omega) = d\tilde{\omega},$$

thus proving that $c_1(E) = c_1(\Lambda^k E)$.

(ii) Let us first assume that E and F are line bundles. Any connections ∇^E and ∇^F on E and F respectively induce a connection ∇ on $E \otimes F$ defined by

$$\nabla(\sigma^E \otimes \sigma^F) := (\nabla^E \sigma^E) \otimes \sigma^F + \sigma^E \otimes (\nabla^F \sigma^F).$$

The corresponding connection forms are then related by $\omega = \omega^E + \omega^F$, so clearly $R^{\nabla} = d\omega = d(\omega^E + \omega^F) = R^{\nabla^E} + R^{\nabla^F}$.

In the general case, if e and f denote the ranks of E and F, the canonical isomorphism $\Lambda^{ef}(E \otimes F) \cong (\Lambda^e E)^{\otimes f} \otimes (\Lambda^f F)^{\otimes e}$ together with (i) yields

$$c_1(E \otimes F) = c_1(\Lambda^{ef}(E \otimes F)) = fc_1(\Lambda^e E) + ec_1(\Lambda^f F) = fc_1(E) + ec_1(F).$$

(iii) Again, since $(\Lambda^k E)^*$ is isomorphic to $\Lambda^k(E^*)$, we can suppose that E is a line bundle. But in this case the canonical isomorphism $E \otimes E^* \simeq \mathbb{C}$ (where $\mathbb{C}$ denotes the trivial line bundle) shows that $0 = c_1(\mathbb{C}) = c_1(E \otimes E^*) = c_1(E) + c_1(E^*)$. $\square$

16.3. Exercises

(1) Consider the change of variables $z = r\cos\theta + ir\sin\theta$. Show that for every function $f : U \subset \mathbb{C} \to \mathbb{C}$, the following formula holds:

$$\bar{\partial}\partial f = \frac{i}{2}\left(r\frac{\partial^2 f}{\partial r^2} + \frac{1}{r}\frac{\partial^2 f}{\partial \theta^2} + \frac{\partial f}{\partial r} \right) dr \wedge d\theta.$$

(2) Show that the first Chern class of the trivial bundle vanishes.

(3) Show that if E is a complex line bundle, there is a canonical isomorphism $E \otimes E^* \simeq \mathbb{C}$.

(4) Let ∇ be any connection on a complex bundle E and let ∇^* be the induced connection on the dual E^* of E defined by

$$(\nabla_X^* \sigma^*)(\sigma) := \partial_X(\sigma^*(\sigma)) - \sigma^*(\nabla_X\sigma).$$

Show that

$$R^{\nabla^*}(X,Y) = -(R^\nabla(X,Y))^*, \tag{16.5}$$

where $A^* \in \mathrm{End}(E^*)$ denotes the adjoint of $A \in \mathrm{End}(E)$, defined by

$$A^*(\sigma^*)(\sigma) := \sigma^*(A(\sigma)).$$

CHAPTER 17

The Ricci form of Kähler manifolds

17.1. Kähler metrics as geometric U_m-structures

We start with a short review on G-structures which will help us to characterize Kähler and Ricci-flat Kähler metrics. Let M be an n-dimensional manifold and let G be any closed subgroup of $\mathrm{Gl}_n(\mathbb{R})$.

DEFINITION 17.1. *A topological G-structure on M is a reduction of the principal frame bundle* $\mathrm{Gl}(M)$ *to G. A geometrical G-structure is given by a topological G-structure* $G(M)$ *together with a torsion-free connection on* $G(M)$.

Let us give some examples. An *orientation* on M is a $\mathrm{Gl}_n^+(\mathbb{R})$-structure. An almost complex structure is a $\mathrm{Gl}_m(\mathbb{C})$-structure, for $n = 2m$. A Riemannian metric is an O_n-structure. Recall (Proposition 4.7) that if the group G is the stabilizer of an element ξ of some representation $\rho : \mathrm{Gl}_n(\mathbb{R}) \to \mathrm{End}(V)$, then a G-structure is simply a section σ in the associated bundle $\mathrm{Gl}(M) \times_\rho \mathcal{O}$, where $\mathcal{O}$ denotes the $\mathrm{Gl}_n(\mathbb{R})$-orbit of ξ in V. By Theorem 5.16, the G-structure is geometrical if and only if there exists a torsion-free linear connection on M with respect to which σ is parallel.

PROPOSITION 17.2. *The U_m-structure defined by an almost complex structure J together with a Hermitian metric h on a manifold M is geometrical if and only if the metric is Kähler.*

PROOF. The point here is that if G is a closed subgroup of O_n then there exists *at most* one torsion-free connection on any G-structure (by the uniqueness of the Levi–Civita connection). As $U_m = O_{2m} \cap \mathrm{Gl}_m(\mathbb{C})$, the U_m-structure is geometrical if and only if the tensor defining it (namely J) is parallel with respect to the Levi–Civita connection, which by Theorem 11.5 just means that h is Kähler. $\square$

17.2. The Ricci form as curvature form on the canonical bundle

We now turn back to our main objects of interest. Let (M^{2m}, h, J) be a Kähler manifold with Ricci form ρ and canonical bundle $K := \Lambda^{m,0}M$. As before, we will interpret the tangent bundle TM as a complex (actually holomorphic) Hermitian vector bundle over M, where the multiplication by

i corresponds to the tensor J and the Hermitian structure is $h - i\Omega$. From Proposition 11.8 we know that the Levi–Civita connection ∇ on M coincides with the Chern connection on TM.

LEMMA 17.3. *The curvature operator $R^\nabla \in \Gamma(\Lambda^2 M \otimes \mathrm{End}(TM))$ of the Chern connection and the curvature tensor R of the Levi–Civita connection are related by*

$$R^\nabla_{X,Y}(\xi) = R(X,Y)\xi,$$

where X, Y are vector fields on M and ξ is a section of TM.

PROOF. The proof is a simple play with definitions. Let $\{e_i\}$ denote a local basis of vector fields on M and let $\{e_i^*\}$ denote the dual local basis of $\Lambda^1 M$. Then by (10.1)

$$R^\nabla \xi = \nabla^2 \xi = \sum_{i=1}^{2m} \nabla(e_i^* \otimes \nabla_{e_i}\xi) = \sum_{i=1}^{2m} de_i^* \otimes \nabla_{e_i}\xi - \sum_{i,j=1}^{2m} e_i^* \wedge e_j^* \otimes \nabla_{e_j}\nabla_{e_i}\xi.$$

Denoting $X_i := e_i^*(X)$ and $Y_i := e_i^*(Y)$ and using again the summation convention on repeating subscripts we then obtain

$$
\begin{aligned}
R^\nabla_{X,Y}(\xi) &= de_i^*(X,Y)\nabla_{e_i}\xi - (e_i^* \wedge e_j^*)(X,Y)\nabla_{e_j}\nabla_{e_i}\xi \\
&= (\partial_X(Y_i) - \partial_Y(X_i) - e_i^*([X,Y]))\nabla_{e_i}\xi - (X_iY_j - X_jY_i)\nabla_{e_j}\nabla_{e_i}\xi \\
&= -\nabla_{[X,Y]}\xi + (\partial_X(Y_i) - \partial_Y(X_i))\nabla_{e_i}\xi - X_i\nabla_Y\nabla_{e_i}\xi + Y_i\nabla_X\nabla_{e_i}\xi \\
&= -\nabla_{[X,Y]}\xi - \nabla_Y\nabla_X\xi + \nabla_X\nabla_Y\xi = R(X,Y)\xi.
\end{aligned}
$$

$\square$

We are now ready to prove the following characterization of the Ricci form ρ on Kähler manifolds:

PROPOSITION 17.4. *The curvature of the Chern connection of the canonical line bundle is equal to $i\rho$ acting by scalar multiplication.*

PROOF. Since we will need to distinguish between complex and real traces below, we will make this explicit by a superscript. We first fix some notations: let r and r^* be the curvatures of the Chern connections of $K := \Lambda^{m,0}M$ and $K^* := \Lambda^{0,m}M$. We have proved in an exercise (see (16.5)) that they are related by $r = -r^*$. The Hermitian structure H on TM induces a Hermitian structure, also denoted by H, on its maximal exterior power $\Lambda^m(TM)$, and the connection on $\Lambda^m(TM)$ induced by the Chern connection of (TM, H) is clearly the Chern connection of $(\Lambda^m(TM), H)$. Since $\Lambda^m(TM)$ is isomorphic to K^*, the proof of Proposition 16.4 together with Lemma 17.3 yields

$$r^*(X,Y) = \mathrm{Tr}(R^\nabla(X,Y)) = \mathrm{Tr}(R^g(X,Y)).$$

By Proposition 12.2 we then obtain

$$
\begin{aligned}
i\rho(X,Y) &= i\mathrm{Ric}(JX,Y) = \frac{i}{2}\mathrm{Tr}^{\mathbb{R}}(R^g(X,Y)\circ J) \\
&= \frac{i}{2}(2i\mathrm{Tr}^{\mathbb{C}}(R^g(X,Y))) = -\mathrm{Tr}^{\mathbb{C}}(R^g(X,Y)) \\
&= -r^*(X,Y) = r(X,Y),
\end{aligned}
$$

where we used the fact that

$$
\mathrm{Tr}^{\mathbb{R}}(A^{\mathbb{R}}\circ J) = 2i\mathrm{Tr}^{\mathbb{C}}(A)
$$

for every skew-Hermitian endomorphism A. $\qquad\square$

17.3. Ricci-flat Kähler manifolds

Let (M^{2m}, h, J) be a Kähler manifold with canonical bundle K (endowed with the Hermitian structure induced from the Kähler metric on TM) and Ricci form ρ. We suppose, for simplicity, that M is simply connected.

THEOREM 17.5. *The five statements below are equivalent:*

(1) *M is Ricci-flat.*
(2) *The Chern connection of the canonical bundle K is flat.*
(3) *There exists a ∇-parallel complex volume form, that is, a parallel section of $\Lambda^{m,0}M$.*
(4) *M has a geometrical SU_m-structure.*
(5) *The Riemannian holonomy group of M is a subgroup of SU_m.*

If M is not simply connected, one has to replace M by "simply connected neighbourhoods of every point in M" in statements (3) and (4), and the Riemannian holonomy group by "restricted Riemannian holonomy group" in the last statement.

PROOF. (1) $\Longleftrightarrow$ (2) is a direct consequence of Proposition 17.4.

(2) $\Longleftrightarrow$ (3) follows from the general principle that a connection on a line bundle is flat if and only if there exists a parallel section (globally defined if $\pi_1(M) = 0$, and locally defined otherwise).

(3) $\Longleftrightarrow$ (4). The special unitary group SU_m can be defined as the stabilizer of a vector in the canonical representation of U_m on $\Lambda^{m,0}\mathbb{C}$. By Theorem 5.16, there exists a parallel section in $\Lambda^{m,0}M$ if and only if the geometrical U_m-structure defined by the Kähler metric can be further reduced to a geometrical SU_m-structure.

(4) $\Longrightarrow$ (5). If $G(M)$ is a G-structure, the holonomy of a connection in $G(M)$ is contained in G. Now, if M has a geometrical SU_m-structure, the torsion-free connection defining it is just the Levi–Civita connection, therefore the Riemannian holonomy group is a subgroup of SU_m.

(5) $\Longrightarrow$ (4). Corollary 5.15 shows that for every fixed frame u, the Levi–Civita connection can be restricted to the holonomy bundle $P(u)$ (which is a $\mathrm{Hol}(u)$-principal bundle). Thus, if the Riemannian holonomy $\mathrm{Hol}(M)$ of M is a subgroup of SU_m, we get a geometrical SU_m-structure simply by extending the holonomy bundle to SU_m. $\qquad\square$

Notice that by Theorem 16.3 and Proposition 17.4, for a given Kähler manifold (M, h, J), the vanishing of the first Chern class of (M, J) is a necessary condition for the existence of a Ricci-flat Kähler metric on M compatible with J. The converse statement is also true if M is compact, and will be treated in the next chapter.

17.4. Exercises

(1) Let G be a closed subgroup of $\mathrm{Gl}_n(\mathbb{R})$ containing SO_n. Show that every G-structure is geometrical.

(2) Let M^n be a connected differentiable manifold. Prove that M is orientable if and only if its frame bundle $\mathrm{Gl}_n(M)$ is not connected.

(3) Show that a U_m-structure on M defines an almost complex structure together with a Hermitian metric.

(4) Show that a geometrical $\mathrm{Gl}_m(\mathbb{C})$-structure is the same as an integrable almost complex structure. *Hint:* Start with a torsion-free connection ∇ and consider the connection $\tilde\nabla$ defined by $\tilde\nabla_X Y := \nabla_X Y - A_X Y$, where $A_X Y = \frac{1}{4}(2J(\nabla_X J)Y + (\nabla_{JY} J)X + J(\nabla_Y J)X)$. Use the proof of Lemma 11.4 to check that A is symmetric if and only if J is integrable.

(5) Let A be a skew-Hermitian endomorphism of $\mathbb{C}^m$ and let $A^{\mathbb{R}}$ be the corresponding real endomorphism of $\mathbb{R}^{2m}$. Show that
$$\mathrm{Tr}^{\mathbb{R}}(A^{\mathbb{R}} \circ J) = 2i\,\mathrm{Tr}^{\mathbb{C}}(A).$$

(6) The special unitary group SU_m is usually defined as the subgroup of $\mathrm{U}_m \subset \mathrm{Gl}_m(\mathbb{C})$ consisting of complex unitary matrices of determinant 1. Prove that SU_m is equal to the stabilizer in U_m of the form $dz_1 \wedge \cdots \wedge dz_m$.

(7) Let (L, h) be a complex line bundle with Hermitian structure over some smooth manifold M. Prove that the space of Hermitian connections is an affine space over the real vector space $\Omega^1 M$. Equivalently,

there is a free transitive group action of $\Omega^1 M$ on the space of Hermitian connections on L.

(8) If L is a complex line bundle over M, show that every real closed 2-form in the cohomology class $c_1(L) \in H^2(M, \mathbb{R})$ is $i/2\pi$ times the curvature of some connection on L.

CHAPTER 18

The Calabi–Yau theorem

18.1. An overview

We have seen that the first Chern class of any compact Kähler manifold is represented by $(1/2\pi)\rho$. Conversely, we have the following famous result conjectured in the 1950s by Calabi and proved two decades later by Yau:

THEOREM 18.1. (Calabi, Yau) *Let M^m be a compact Kähler manifold with Kähler form Ω and Ricci form ρ. Then for every closed real $(1,1)$-form ρ_1 in the cohomology class $2\pi c_1(M)$, there exists a unique Kähler metric with Kähler form Ω_1 in the same cohomology class as Ω, whose Ricci form is exactly ρ_1. In particular, if the first Chern class of a compact Kähler manifold vanishes, then M carries a Ricci-flat Kähler metric.*

The first step in the proof of Theorem 18.1 is to reformulate the problem in order to reduce it to a so-called *Monge–Ampère equation*. Let us denote by $\mathcal{K}$ the set of Kähler metrics in the same cohomology class as Ω. The global $i\partial\bar\partial$-lemma (Exercise (2), Chapter 15) shows that

$$\mathcal{K} = \left\{ u \in \mathcal{C}^\infty(M) \mid \Omega + i\partial\bar\partial u > 0, \ \int_M u\Omega^m = 0 \right\}, \qquad (18.1)$$

where a real $(1,1)$-form φ is called *positive* if the symmetric tensor $\varphi(\cdot, J\cdot)$ is positive definite (the last condition is needed since u would otherwise be defined only up to a constant).

Now, if g and g_1 are Kähler metrics with Kähler forms Ω and Ω_1 in the same cohomology class, we denote by $dv := (1/m!)\Omega^m$ and $dv_1 := (1/m!)\Omega_1^m$ their volume forms and consider the real function f defined by $e^f\, dv = dv_1$. Since $[\Omega] = [\Omega_1]$ we also have $[\Omega^m] = [\Omega_1^m]$, so by the Stokes theorem,

$$\int_M e^f\, dv = \int_M dv. \qquad (18.2)$$

Let ρ and ρ_1 denote the corresponding Ricci forms. Since $i\rho$ is the curvature of the canonical bundle K_M, for every local holomorphic section ω of K_M we have

$$i\rho = \bar\partial\partial \log g(\omega,\bar\omega) \qquad \text{and} \qquad i\rho_1 = \bar\partial\partial \log g_1(\omega,\bar\omega). \qquad (18.3)$$

It is easy to check that the Hodge $*$-operator acts on $\Lambda^{m,0}$ simply by scalar multiplication with $\varepsilon := i^m(-1)^{\frac{m(m+1)}{2}}$. We thus have

$$\varepsilon\omega \wedge \bar{\omega} = \omega \wedge *\bar{\omega} = g(\omega, \bar{\omega})dv \tag{18.4}$$

and similarly

$$\varepsilon\omega \wedge \bar{\omega} = g_1(\omega, \bar{\omega})dv_1 = e^f g_1(\omega, \bar{\omega})dv. \tag{18.5}$$

From (18.3)–(18.5) we get

$$i\rho_1 - i\rho = \partial\bar{\partial}f. \tag{18.6}$$

This shows that the Ricci form of the Kähler metric $\Omega_1 = \Omega + i\partial\bar{\partial}u$ can be computed by the formula

$$\rho_1 = \rho - i\partial\bar{\partial}f, \qquad \text{where } f = \log\frac{(\Omega + i\partial\bar{\partial}u)^m}{\Omega^m}. \tag{18.7}$$

Now given closed real $(1,1)$-form ρ_1 in the cohomology class $2\pi c_1(M)$, the global $i\partial\bar{\partial}$-lemma (Exercise (2), Chapter 15) shows that there exists some real function f such that $\rho_1 = \rho - i\partial\bar{\partial}f$. Moreover, f is unique if we impose the normalization condition (18.2). We denote by $\mathcal{K}'$ the space of smooth functions on M satisfying this condition. Theorem 18.1 can thus be reformulated as follows:

Theorem 18.2. *The mapping* $Cal : \mathcal{K} \to \mathcal{K}'$ *defined by*

$$Cal(u) = \log\frac{(\Omega + i\partial\bar{\partial}u)^m}{\Omega^m}$$

is a diffeomorphism.

We first show that Cal is injective. It is clearly enough to show that $Cal(u) = 0$ and $u \in \mathcal{K}$ implies $u = 0$. If $Cal(u) = 0$ we have $\Omega_1^m = \Omega^m$, and since 2-forms commute we obtain

$$0 = \Omega_1^m - \Omega^m = i\partial\bar{\partial}u \wedge \sum_{k=0}^{m-1}\Omega_1^k \wedge \Omega^{m-k-1}.$$

Using the formula $2i\partial\bar{\partial} = dd^c$ and the fact that Ω and Ω_1 are closed forms we get after multiplication by u

$$\begin{aligned}
0 &= 2iu\partial\bar{\partial}u \wedge \sum_{k=0}^{m-1}\Omega_1^k \wedge \Omega^{m-k-1} = udd^cu \wedge \sum_{k=0}^{m-1}\Omega_1^k \wedge \Omega^{m-k-1}\\
&= d\left(ud^cu \wedge \sum_{k=0}^{m-1}\Omega_1^k \wedge \Omega^{m-k-1}\right) - du \wedge d^cu \wedge \sum_{k=0}^{m-1}\Omega_1^k \wedge \Omega^{m-k-1}.
\end{aligned}$$

Integrating over M and using the Stokes theorem yields

$$0 = \sum_{k=0}^{m-1}\int_M du \wedge Jdu \wedge \Omega_1^k \wedge \Omega^{m-k-1}. \tag{18.8}$$

Now, since Ω_1 defines a Kähler metric, the tangent bundle of M carries a local basis $\{e_1, Je_1, \ldots, e_m, Je_m\}$ orthonormal with respect to g such that

$$\Omega = \sum_{j=1}^{m} e_j \wedge Je_j \quad \text{and} \quad \Omega_1 = \sum_{j=1}^{m} a_j e_j \wedge Je_j,$$

where a_j are strictly positive local functions. One then gets for every k

$$\Omega_1^k \wedge \Omega^{m-k-1} = *\left(\sum_{j=1}^{m} b_j^k e_j \wedge Je_j\right),$$

for some strictly positive b_j^k. In fact one can compute explicitly

$$b_j^k = k!(m-k-1)! \sum_{\substack{j_1 \neq j, \ldots, j_k \neq j \\ j_1 < \cdots < j_k}} a_{j_1} \ldots a_{j_k}.$$

This shows that the integrand in (18.8) is strictly positive unless $du = 0$. Thus u is a constant, so $u = 0$ because the integral of $u\, dv$ over M vanishes. Therefore Cal is injective.

To prove that it is a local diffeomorphism, we compute its differential at some $u \in \mathcal{K}$. By changing the reference metric if necessary, we may suppose without loss of generality that $u = 0$. For $v \in T_0\mathcal{K}$ we compute

$$\begin{aligned}
Cal_*(v) &= \frac{d}{dt}\Big|_{t=0}(Cal(tv)) = \frac{d}{dt}\Big|_{t=0} \log\left(\frac{(\Omega + it\partial\bar\partial v)^m}{\Omega^m}\right) \\
&= m\frac{i\partial\bar\partial v \wedge \Omega^{m-1}}{\Omega^m} = \Lambda(i\partial\bar\partial v) = -\bar\partial^*\bar\partial v = -\Delta^{\bar\partial} v.
\end{aligned}$$

From the general elliptic theory we know that the Laplace operator is a bijection of the space of functions with zero integral over M. Thus Cal_* is bijective, so the Inverse Function Theorem shows that Cal is a local diffeomorphism.

The surjectivity of Cal, which is the hard part of the theorem, follows from a priori estimates, which show that Cal is proper. We refer the reader to [7], pp. 98–120 for details.

18.2. Exercises

(1) Show that $*\omega = i^{m(m+2)}\omega$ for all $(m, 0)$-forms $\omega \in \Lambda^{m,0}M$ of an almost Hermitian manifold (M^{2m}, h, J).

(2) Prove that the mapping

$$u \mapsto \Omega + i\partial\bar\partial u$$

is indeed a bijection from the set defined in (18.1) to the set of Kähler metrics with Kähler form in the cohomology class $[\Omega]$.

(3) Prove that the total volume of a Kähler metric on a compact manifold depends on only the cohomology class of its Kähler form.

(4) Show that $*(\Omega^{m-1}) = (m-1)!\,\Omega$ on every almost Hermitian manifold M of complex dimension m. Using this, prove that if (M, g, J) is Kähler, then $2\rho \wedge \Omega^{m-1} = (m-1)!\,S\,dv$, where ρ denotes the Ricci form and S is the scalar curvature of g.

(5) Let (M^{2m}, J) be a compact complex manifold. Show that the integral over M of the scalar curvature of a Kähler metric depends on only the cohomology class of its Kähler form Ω and on $c_1(M)$. More precisely one has

$$\int_M S\,dv = \frac{4\pi}{(m-1)!} c_1(M) \cup [\Omega]^{m-1}.$$

CHAPTER 19

Kähler–Einstein metrics

19.1. The Aubin–Yau theorem

We turn our attention to compact Kähler manifolds (M, g) satisfying the Einstein condition

$$\mathrm{Ric} = \lambda g, \qquad \lambda \in \mathbb{R}.$$

We will exclude the case $\lambda = 0$ which was discussed in the previous chapter. If we rescale the metric by a positive constant, the curvature tensor does not change, so neither does the Ricci tensor, which was defined as a trace. This shows that we may suppose that $\lambda = \varepsilon = \pm 1$. The Kähler–Einstein condition reads

$$\rho = \varepsilon \Omega, \qquad \varepsilon = \pm 1.$$

As the first Chern class of M is represented by $\rho/2\pi$, we see that a necessary condition for the existence of a Kähler–Einstein manifold on a given compact Kähler manifold is that its first Chern class is *definite* (positive or negative), in the sense that it has a positive or negative $(1, 1)$ representative (see Definition 21.1 below). In the negative case, this condition turns out to be also sufficient:

THEOREM 19.1. (Aubin, Yau) *A compact complex manifold with negative first Chern class admits a unique Kähler–Einstein metric with Einstein constant $\varepsilon = -1$.*

We will treat simultaneously the two cases $\varepsilon = \pm 1$, in order to emphasize the difficulties that show up in the case $\varepsilon = 1$.

As before, we first reformulate the problem. Let (M^{2m}, J) be a compact complex manifold with definite first Chern class $c_1(M)$. By definition, there exists a positive closed $(1, 1)$-form Ω representing the cohomology class $2\pi\varepsilon c_1(M)$. Let $g := \Omega(\cdot, J\cdot)$ be the Kähler metric defined by Ω, and let ρ denote its Ricci form. Then $[\Omega] = 2\pi\varepsilon c_1(M) = [\varepsilon\rho]$, so the global $i\partial\bar{\partial}$-lemma shows that there exists some function f with

$$\rho = \varepsilon\Omega + i\partial\bar{\partial}f. \tag{19.1}$$

We are looking for a new Kähler metric g_1 with Kähler form Ω_1 and Ricci form ρ_1 such that $\rho_1 = \varepsilon\Omega_1$. Suppose we have such a metric. From our choice for Ω we have

$$[2\pi\Omega] = \varepsilon c_1(M) = [2\pi\varepsilon\rho_1] = [2\pi\Omega_1].$$

From this equation and the global $i\partial\bar{\partial}$-lemma it is clear that there exists a unique function $u \in \mathcal{K}$ such that $\Omega_1 = \Omega + i\partial\bar{\partial}u$. Now the previously obtained formula (18.7) for the Ricci form of the new metric reads

$$\rho_1 = \rho - i\partial\bar{\partial}\log\frac{(\Omega + i\partial\bar{\partial}u)^m}{\Omega^m}. \tag{19.2}$$

Using (19.1) and (19.2), the Kähler–Einstein condition for g_1 becomes

$$\varepsilon\Omega + i\partial\bar{\partial}f - i\partial\bar{\partial}\log\frac{(\Omega + i\partial\bar{\partial}u)^m}{\Omega^m} = \varepsilon\Omega_1, \tag{19.3}$$

which is equivalent to

$$\log\frac{(\Omega + i\partial\bar{\partial}u)^m}{\Omega^m} + \varepsilon u = f + const. \tag{19.4}$$

Conversely, if $u \in \mathcal{C}_+^\infty(M)$ satisfies this equation, then the Kähler metric $\Omega_1 := \Omega + i\partial\bar{\partial}u$ is Kähler–Einstein (we denote by $\mathcal{C}_+^\infty(M)$ the space of all smooth functions u on M such that $\Omega + i\partial\bar{\partial}u > 0$). The Aubin–Yau theorem is therefore equivalent to the fact that the mapping

$$Cal^\varepsilon : \mathcal{C}_+^\infty(M) \to \mathcal{C}^\infty(M), \qquad Cal^\varepsilon(u) := Cal(u) + \varepsilon u$$

is a diffeomorphism.

The injectivity of Cal^- can be proved as follows. Suppose that $Cal^-(u_1) = Cal^-(u_2)$ and denote $\Omega_1 := \Omega + i\partial\bar{\partial}u_1$ and $\Omega_2 := \Omega + i\partial\bar{\partial}u_2$. Then

$$\log\frac{\Omega_1^m}{\Omega^m} - u_1 = \log\frac{\Omega_2^m}{\Omega^m} - u_2,$$

hence, denoting the difference $u_2 - u_1$ by u:

$$\log\frac{(\Omega_1 + i\partial\bar{\partial}u)^m}{\Omega_1^m} = u. \tag{19.5}$$

At a point where u attains its maximum, the $(1,1)$-form $i\partial\bar{\partial}u$ is negative semi-definite, since we can write (for any vector X parallel at that point)

$$i\partial\bar{\partial}u(X, JX) = \frac{1}{2}(dd^c u)(X, JX) = \frac{1}{2}(\partial_X(d^c u(JX)) - \partial_{JX}(d^c u(X)))$$

$$= \frac{1}{2}(H^u(X, X) + H^u(JX, JX)) \leq 0,$$

the Hessian H^u of u being of course negative semi-definite at a point where u reaches its maximum. Taking into account (19.5) we see that $u \leq 0$ at each of its maximum points, so $u \leq 0$ on M. Similarly, $u \geq 0$ at each minimum point, so finally $u = 0$ on M, thus proving the injectivity of Cal^-.

We have already computed the differential of Cal at $u = 0$ applied to some $v \in T_0 \mathcal{C}_+^\infty(M)$:

$$Cal_*(v) = -\Delta^{\bar{\partial}} v.$$

Consequently

$$Cal_*^-(v) = -v - \Delta^{\bar{\partial}} v$$

is a bijection of $\mathcal{C}^\infty(M)$ since the self-adjoint elliptic operator $v \mapsto \frac{1}{2}\Delta v + v$ has obviously no kernel and its index is zero. Of course, this argument fails for Cal^+.

As before, the surjectivity of Cal^- is harder to prove and requires non-trivial analysis (see [1], p. 329).

19.2. Holomorphic vector fields on Kähler–Einstein manifolds

Let (M^{2m}, g, J) be a compact Kähler manifold. We start by showing the following:

LEMMA 19.2. *Let ξ be a holomorphic (real) vector field with dual 1-form also denoted by ξ. Then ξ can be decomposed in a unique manner as*

$$\xi = df + d^c h + \xi^H,$$

where f and h are functions with vanishing integral and ξ^H is the harmonic part of ξ in the usual Hodge decomposition.

PROOF. Since ξ is holomorphic we have $\mathcal{L}_\xi J = 0$, so $[\xi, JX] = J[\xi, X]$ for every vector field X. Thus $\nabla_{JX}\xi = J\nabla_X\xi$, so taking the scalar product with some vector field JY and skew-symmetrizing yields $d\xi(JX, JY) = d\xi(X, Y)$, i.e. $d\xi$ is of type $(1, 1)$. The global dd^c-lemma shows that $d\xi = dd^c h$ for some function h. The form $\xi - d^c h$ is closed, so the Hodge decomposition theorem yields

$$\xi - d^c h = df + \xi_0$$

for some function f and some harmonic 1-form ξ_0. Comparing this formula with the Hodge decomposition of ξ and using the fact that harmonic 1-forms are L^2-orthogonal to $d\mathcal{C}^\infty(M)$, $d^c\mathcal{C}^\infty(M)$ and $\delta\Omega^2(M)$, shows that ξ_0 equals ξ^H, the harmonic part of ξ. Finally, the uniqueness of f and h follows easily from the normalization condition, together with the fact that $d\mathcal{C}^\infty(M)$ and $d^c\mathcal{C}^\infty(M)$ are L^2-orthogonal. $\square$

Next, we have the following characterization of real holomorphic and Killing vector fields on compact Kähler–Einstein manifolds with positive scalar curvature.

LEMMA 19.3. *A vector field ξ (resp. ζ) on a compact Kähler–Einstein manifold M^{2m} with positive scalar curvature S is Killing (resp. holomorphic)* *if and only if $\xi = Jdh$ (resp. $\zeta = df + d^c h$) where h (resp. f and h) are eigenfunctions of the Laplace operator corresponding to the eigenvalue S/m.*

PROOF. Since (M, g) is Einstein, the Ricci tensor of M, viewed as a field of endomorphisms via the metric g satisfies $\mathrm{Ric}(X) = (S/2m)X$ for every vector X. Let ξ be a vector field on M. If we view as usual TM as a holomorphic vector bundle, then the Weitzenböck formula (see (20.10) below) yields

$$2\bar{\partial}^* \bar{\partial} \xi = \nabla^* \nabla \xi + i \rho \xi = \nabla^* \nabla \xi - \mathrm{Ric}(\xi) = \nabla^* \nabla \xi - \frac{S}{2m} \xi. \qquad (19.6)$$

The Bochner formula (Exercise (3) in the next chapter) reads

$$\Delta \xi = \nabla^* \nabla \xi + \mathrm{Ric}(\xi) = \nabla^* \nabla \xi + \frac{S}{2m} \xi. \qquad (19.7)$$

Since $S > 0$, this shows that there are no harmonic 1-forms on M.

Suppose that ζ is holomorphic. From Lemma 19.2, ζ can be written as a sum $\zeta = df + d^c h$, where f and h have vanishing integrals over M. Now, subtracting (19.6) from (19.7) yields $\Delta \zeta = (S/m)\zeta$, so $\Delta(df + d^c h) = d((S/m)f) + d^c((S/m)h)$, and since Δ commutes with d and d^c, and the images of d and d^c are L^2-orthogonal, this yields $\Delta f = (S/m)f + c_1$ and $\Delta h = (S/m)h + c_2$. Finally the constants c_1 and c_2 have to vanish because of the normalization condition.

If ξ is a Killing vector field, then it is in particular holomorphic by Proposition 15.5. The previous argument shows that $\xi = df + d^c h$ for some eigenfunctions f, h of Δ. The codifferential of every Killing vector field vanishes, and moreover δ anti-commutes with d^c. Thus $0 = \delta \xi = \delta df$, showing that $df = 0$, so $\xi = d^c h$ with $\Delta h = (S/m)h$.

Conversely, suppose that $\xi = df + d^c h$ and f and h are eigenfunctions of the Laplace operator corresponding to the eigenvalue S/m. Then $\Delta \xi = (S/m)\xi$, so from (19.7) we get

$$\frac{S}{m} \xi = \nabla^* \nabla \xi + \frac{S}{2m} \xi.$$

Then (19.6) shows that ξ is holomorphic.

If moreover $df = 0$, we have

$$\mathcal{L}_\xi \Omega = d(\xi \lrcorner \Omega) + \xi \lrcorner d\Omega = d(J\xi) = -ddh = 0,$$

where Ω is the Kähler form of M. Together with $\mathcal{L}_\xi J = 0$, this shows that $\mathcal{L}_\xi g = 0$, so ξ is Killing. $\qquad \square$

We are now ready to prove the following result of Matsushima:

THEOREM 19.4. *The Lie algebra $\mathbf{g}(M)$ of Killing vector fields on a compact Kähler–Einstein manifold M with positive scalar curvature is a real form* *of the Lie algebra $\mathfrak{h}(M)$ of (real) holomorphic vector fields on M. In particular $\mathfrak{h}(M)$ is reductive, i.e. it is the direct sum of its centre and a semi-simple Lie algebra.*

PROOF. Let $F : \mathfrak{g}(M) \otimes \mathbb{C} \to \mathfrak{h}(M)$ be the linear map given by $F(\xi + i\zeta) :=$ $\xi + J\zeta$. Since J maps holomorphic vector fields to holomorphic vector fields, F is well-defined. The two lemmas above clearly show that F is a vector space isomorphism. Moreover, F is a Lie algebra morphism because Killing vector fields are holomorphic:

$$
\begin{aligned}
F([\xi + i\zeta, \xi_1 + i\zeta_1]) &= [\xi, \xi_1] - [\zeta, \zeta_1] + J([\xi, \zeta_1] + [\zeta, \xi_1]) \\
&= [\xi, \xi_1] + J^2[\zeta, \zeta_1] + [J\xi, \zeta_1] + [J\zeta, \xi_1] \\
&= [\xi, \xi_1] + [J\zeta, J\zeta_1] + [J\xi, \zeta_1] + [J\zeta, \xi_1] \\
&= [F(\xi + i\zeta), F(\xi_1 + i\zeta_1)].
\end{aligned}
$$

The last statement of the theorem follows from the fact that the isometry group of M is compact, and every Lie algebra of compact type, as well as its complexification, is reductive. $\square$

There exist compact Kähler manifolds with positive first Chern class whose Lie algebra of holomorphic vector fields is not reductive, like for instance the blow-up of $\mathbb{C}P^2$ at one point (see [1], p. 331). Therefore such manifolds carry no Kähler–Einstein metrics, showing that Theorem 19.1 can not hold in the positive case.

The Matsushima theorem was generalized to constant scalar curvature Kähler metrics by Lichnerowicz and to *extremal* Kähler metrics by Calabi. Details can be found in the recent monograph [2] by P. Gauduchon, which contains a very up to date state of research in this field.

19.3. Exercises

(1) Let η be the dual 1-form to a Killing vector field. Show that the codifferential of η vanishes.

(2) With the previous notations, show that if $d\eta = 0$ then ξ is parallel with respect to the Levi–Civita connection.

(3) Let ξ be a Killing vector field on a compact Kähler manifold (M, g, J) such that $J\xi$ is also Killing. Prove that ξ is then parallel with respect to the Levi–Civita connection. *Hint:* Compute the Lie derivative of the Kähler form with respect to $J\xi$ using the Cartan formula (Theorem 3.3) and apply the previous exercise, taking into account Proposition 15.5.

(4) Prove that every Killing vector field on a compact Kähler–Einstein manifold with positive scalar curvature has at least two zeros.

CHAPTER 20

Weitzenböck techniques

20.1. The Weitzenböck formula

The aim of the next chapters is to derive vanishing results under certain positivity assumptions on the curvature using Weitzenböck techniques.

The general principle is the following: let $(E, h) \to M$ be some holomorphic Hermitian bundle over a compact Kähler manifold (M^{2m}, g, J), with holomorphic structure denoted $\bar{\partial} : \Omega^{p,q}(E) \to \Omega^{p,q+1}(E)$ and Chern connection denoted $\nabla : \Omega^{p,q}(E) \to \Gamma(\Lambda^1_{\mathbb{C}} M \otimes \Lambda^{p,q}(E))$. If $\bar{\partial}^*$ and ∇^* are the formal adjoints of $\bar{\partial}$ and ∇, it turns out that the difference of the differential operators of order two $\nabla^* \nabla$ and $2(\bar{\partial}^* \bar{\partial} + \bar{\partial} \bar{\partial}^*)$ acting on $\Omega^{p,q}(E)$ is a zero-order operator, depending only on the curvature of the Chern connection:

$$2(\bar{\partial}^* \bar{\partial} + \bar{\partial} \bar{\partial}^*) = \nabla^* \nabla + \mathcal{R}, \tag{20.1}$$

where $\mathcal{R}$ is a section of $\mathrm{End}(\Lambda^{p,q}(E))$. If $\mathcal{R}$ is a *positive operator* on $\Lambda^{p,0}(E)$, then every holomorphic section of $\Lambda^{p,0}(E)$ is ∇-parallel, and if $\mathcal{R}$ is *strictly positive* on $\Lambda^{p,0}(E)$, then E has no holomorphic section. This follows by applying (20.1) to some holomorphic section σ of $\Lambda^{p,0}(E)$, taking the scalar product with σ and integrating over M, using the fact that $\bar{\partial}^*$ vanishes identically on $\Omega^{p,0}(E)$.

We start with the following technical lemma:

LEMMA 20.1. *If $\{e_j\}$ is a local orthonormal basis in TM (identified via the metric g with a local orthonormal basis of $\Lambda^1 M$), and ∇ denotes the Chern connection of E, as well as its extension to $\Lambda^{p,q}(E) = \Lambda^{p,q} M \otimes E$ using the Levi–Civita connection on the left hand side of this tensor product, then $\bar{\partial}$, $\bar{\partial}^*$, ∇^* and $\nabla^* \nabla$ are given locally by*

$$\bar{\partial} : \Omega^{p,q}(E) \to \Omega^{p,q+1}(E), \qquad \bar{\partial}\sigma = \frac{1}{2}(e_j - iJe_j) \wedge \nabla_{e_j}(\sigma), \tag{20.2}$$

$$\bar{\partial}^* : \Omega^{p,q}(E) \to \Omega^{p,q-1}(E), \qquad \bar{\partial}^*\sigma = -\frac{1}{2}(e_j + iJe_j) \lrcorner \nabla_{e_j}(\sigma), \tag{20.3}$$

$$\nabla^* : \Gamma(\Lambda^1_{\mathbb{C}} M \otimes \Lambda^{p,q}(E)) \to \Omega^{p,q}(E), \qquad \nabla^*(\omega \otimes \sigma) = (\delta\omega)\sigma - \nabla_\omega \sigma, \tag{20.4}$$

$$\nabla^* \nabla : \Omega^{p,q}(E) \to \Omega^{p,q}(E), \qquad \nabla^* \nabla\sigma = \nabla_{\nabla_{e_j} e_j} \sigma - \nabla_{e_j} \nabla_{e_j} \sigma. \tag{20.5}$$

Notice that here, as well as in the next two chapters, we use again the summation convention on repeating subscripts.

PROOF. If (E, h^E) and (F, h^F) are Hermitian bundles, their tensor product inherits a natural Hermitian structure given by

$$h^{E \otimes F}(\sigma^E \otimes \sigma^F, s^E \otimes s^F) := h^E(\sigma^E, s^E) h^F(\sigma^F, s^F).$$

The Hermitian structure on $\Lambda^{p,q}(E)$ with respect to which one defines the adjoint operators above, is obtained in this way from the Hermitian structure h of E and the Hermitian structure H of $\Lambda^{p,q}M$ given by (14.11). By a slight abuse of notation, we will use the same symbol H for this Hermitian structure on $\Lambda^{p,q}(E)$.

The relation (20.2) is more or less tautological, using the definition of $\bar{\partial}$ and the fact that $e_j - iJe_j$ is a $(1,0)$-vector, identified via the metric g with a $(0,1)$-form. Of course, the wedge product there only concerns the $\Lambda^{p,q}M$ part of σ.

For $\sigma \in \Omega^{p,q}(E)$ and $s \in \Omega^{p,q-1}(E)$ we define the 1-form α by

$$\alpha(X) := \frac{1}{2} H((X + iJX) \lrcorner \sigma, s).$$

By choosing the local basis $\{e_j\}$ parallel at a point for simplicity, we get at that point:

$$
\begin{aligned}
-\delta\alpha &= e_j(\alpha(e_j)) = \frac{1}{2} H((e_j + iJe_j) \lrcorner \nabla_{e_j}\sigma, s) + \frac{1}{2} H((e_j + iJe_j) \lrcorner \sigma, \nabla_{e_j}s) \\
&= \frac{1}{2} H((e_j + iJe_j) \lrcorner \nabla_{e_j}\sigma, s) + \frac{1}{2} H(\sigma, (e_j - iJe_j) \wedge \nabla_{e_j}s) \\
&= \frac{1}{2} H((e_j + iJe_j) \lrcorner \nabla_{e_j}\sigma, s) + H(\sigma, \bar{\partial}s).
\end{aligned}
$$

Lemma 14.3 thus shows that the operator $-\frac{1}{2}(e_j + iJe_j) \lrcorner \nabla_{e_j}$ is the formal adjoint of $\bar{\partial}$.

The proof of (20.4) is similar: for $\omega \otimes \sigma \in \Gamma(\Lambda_{\mathbb{C}}^1 M \otimes \Lambda^{p,q}(E))$ and $s \in \Omega^{p,q}(E)$ we define the 1-form α by

$$\alpha(X) := H((\omega(X))\sigma, s)$$

and compute

$$
\begin{aligned}
-\delta\alpha &= e_j(\alpha(e_j)) = -H((\delta\omega)\sigma, s) + H((\omega(e_j))\nabla_{e_j}\sigma, s) + H((\omega(e_j))\sigma, \nabla_{e_j}s) \\
&= H(\nabla_\omega\sigma - (\delta\omega)\sigma, s) + H(\omega \otimes \sigma, \nabla s),
\end{aligned}
$$

whence $\nabla^*(\omega \otimes \sigma) = (\delta\omega)\sigma - \nabla_\omega\sigma$.

Finally, we apply (20.4) to some section $\nabla\sigma = e_j \otimes \nabla_{e_j}\sigma$ of $\Lambda_{\mathbb{C}}^1 M \otimes \Lambda^{p,q}(E)$ and get

$$
\begin{aligned}
\nabla^*\nabla\sigma &= (\delta e_j)\nabla_{e_j}\sigma - \nabla_{e_j}\nabla_{e_j}\sigma = -g(e_k, \nabla_{e_k}e_j)\nabla_{e_j}\sigma - \nabla_{e_j}\nabla_{e_j}\sigma \\
&= g(\nabla_{e_k}e_k, e_j)\nabla_{e_j}\sigma - \nabla_{e_j}\nabla_{e_j}\sigma = \nabla_{\nabla_{e_j}e_j}\sigma - \nabla_{e_j}\nabla_{e_j}\sigma.
\end{aligned}
$$

$\square$

We are now ready for the main result of this section:

THEOREM 20.2. *Let* $(E, h) \to (M^{2m}, g, J)$ *be a holomorphic Hermitian bundle over a Kähler manifold* M. *For vectors* $X, Y \in TM$, *let* $\tilde{R}(X, Y) \in \mathrm{End}(\Lambda^{p,q}(E))$ *be the curvature operator of the tensor product connection on* $\Lambda^{p,q}(E)$ *induced by the Levi–Civita connection of* $\Lambda^{p,q}M$ *and the Chern connection of* E. *Then the following formula holds*

$$2(\bar{\partial}^*\bar{\partial} + \bar{\partial}\bar{\partial}^*) = \nabla^*\nabla + \mathcal{R}, \tag{20.6}$$

where $\mathcal{R}$ *is the section of* $\mathrm{End}(\Lambda^{p,q}(E))$ *defined by*

$$\mathcal{R}(\sigma) := \frac{i}{2}\tilde{R}(Je_j, e_j)\sigma - \frac{1}{2}(e_j - iJe_j) \wedge (e_k + iJe_k) \lrcorner (\tilde{R}(e_j, e_k)\sigma). \tag{20.7}$$

PROOF. The proof is a simple computation in a local orthonormal frame parallel at a point, using Lemma 20.1, (20.2), (20.3), and (20.5):

$$
\begin{aligned}
2\bar{\partial}^*\bar{\partial} &= -\frac{1}{2}\left((e_k + iJe_k) \lrcorner \nabla_{e_k}((e_j - iJe_j) \wedge \nabla_{e_j})\right) \\
&= -\frac{1}{2}\left((e_k + iJe_k) \lrcorner ((e_j - iJe_j) \wedge \nabla_{e_k}\nabla_{e_j})\right) \\
&= -(g(e_k, e_j) + ig(Je_k, e_j))\nabla_{e_k}\nabla_{e_j} \\
&\quad + \frac{1}{2}\left((e_j - iJe_j) \wedge (e_k + iJe_k) \lrcorner \nabla_{e_k}\nabla_{e_j}\right) \\
&= \nabla^*\nabla - ig(Je_k, e_j)\nabla_{e_k}\nabla_{e_j} \\
&\quad + \frac{1}{2}\left((e_j - iJe_j) \wedge (e_k + iJe_k) \lrcorner \nabla_{e_j}\nabla_{e_k}\right) \\
&\quad + \frac{1}{2}\left((e_j - iJe_j) \wedge (e_k + iJe_k) \lrcorner \tilde{R}(e_k, e_j)\right) \\
&= \nabla^*\nabla - \frac{i}{2}g(Je_k, e_j)\tilde{R}(e_k, e_j) \\
&\quad + \frac{1}{2}\left((e_j - iJe_j) \wedge \nabla_{e_j}((e_k + iJe_k) \lrcorner \nabla_{e_k})\right) \\
&\quad + \frac{1}{2}\left((e_j - iJe_j) \wedge (e_k + iJe_k) \lrcorner \tilde{R}(e_k, e_j)\right) \\
&= \nabla^*\nabla - 2\bar{\partial}\bar{\partial}^* + \mathcal{R}.
\end{aligned}
$$

□

20.2. Vanishing results on Kähler manifolds

Most of the applications will concern the case $q = 0$. The expression of the curvature term becomes then particularly simple, since the last term in

(20.7) automatically vanishes. Let $\rho^{(p)}$ denote the action of the Ricci-form of M on $\Lambda^{p,0}$ given by

$$\rho^{(p)}\omega := \rho(e_j) \wedge e_j \lrcorner \omega,$$

where ρ is regarded here as skew-symmetric endomorphism of TM by the usual formula $g(\rho(X), Y) = \rho(X, Y)$. This action preserves the space $\Lambda^{p,0}$.

PROPOSITION 20.3. If $q = 0$, for every section $\omega \otimes \xi$ of $\Lambda^{p,0}(E)$ we have

$$2\bar{\partial}^*\bar{\partial}(\omega \otimes \xi) = \nabla^*\nabla(\omega \otimes \xi) + i(\rho^{(p)}\omega) \otimes \xi + \frac{i}{2}\omega \otimes R^E(Je_j, e_j)\xi, \quad (20.8)$$

where R^E is the curvature of E.

PROOF. The curvature $\tilde{R}$ of $\Lambda^{p,0}(E)$ decomposes in a sum

$$\tilde{R}(X, Y)(\omega \otimes \xi) = (R(X, Y)\omega) \otimes \xi + \omega \otimes R^E(X, Y)(\xi), \quad (20.9)$$

where R is the Riemannian curvature. It is an easy exercise to check that the Riemannian curvature operator acts on forms by

$$R(X, Y)(\omega) = R(X, Y)e_k \wedge (e_k \lrcorner \omega).$$

From Proposition 12.2(i) we have $2\rho = R(Je_j, e_j)$ as endomorphisms of the tangent space of M. Therefore (20.8) follows from Theorem 20.2 and (20.9). $\square$

We now apply this proposition to obtain several vanishing results.

THEOREM 20.4. Let M be a compact Kähler manifold. If the Ricci curvature of M is negative definite (i.e. $\mathrm{Ric}(X, X) < 0$ for all non-zero $X \in TM$) then M carries no holomorphic vector field.

PROOF. Let us take $p = 0$ and $E = T^{1,0}M$ in Proposition 20.3. If ξ is a holomorphic vector field, we have

$$0 = 2\bar{\partial}^*\bar{\partial}\xi = \nabla^*\nabla\xi + \frac{i}{2}R(Je_j, e_j)\xi = \nabla^*\nabla\xi + i\rho(\xi). \quad (20.10)$$

Taking the (Hermitian) scalar product with ξ in this formula and integrating over M, using the fact that $\rho\xi = \mathrm{Ric}(J\xi) = i\mathrm{Ric}(\xi)$ yields

$$0 = \int_M H(\nabla^*\nabla\xi - \mathrm{Ric}(\xi), \xi)dv = \int_M |\nabla\xi|^2 - H(\mathrm{Ric}(\xi), \xi)dv.$$

Thus, if Ric is negative definite, ξ has to vanish identically. $\square$

THEOREM 20.5. Let M be a compact Kähler manifold. If the Ricci curvature of M vanishes, then every holomorphic form is parallel. If the Ricci curvature of M is positive definite, then there exist no holomorphic $(p, 0)$-forms on M for $p > 0$.

PROOF. We take E to be trivial and apply (20.8) to some holomorphic $(p,0)$-form ω. Since $\rho = 0$ we get $0 = \nabla^*\nabla\omega$. Taking the Hermitian product with ω and integrating over M yields the result.

Suppose now that Ric is positive definite. From (20.8) applied to some holomorphic $(p,0)$-form ω we get

$$0 = \nabla^*\nabla\omega + i\rho^{(p)}(\omega). \tag{20.11}$$

The interior product of a $(0,1)$-vector and ω vanishes, so $JX \lrcorner \omega = iX \lrcorner \omega$ for all $X \in TM$. We thus get

$$\begin{aligned} i\rho^{(p)}(\omega) &= i\rho(e_j) \wedge e_j \lrcorner \omega = i\rho(Je_j) \wedge Je_j \lrcorner \omega = -\rho(Je_j) \wedge e_j \lrcorner \omega \\ &= \mathrm{Ric}(e_j) \wedge e_j \lrcorner \omega. \end{aligned}$$

Since Ric is positive, the endomorphism induced on $(p,0)$-forms by $\omega \mapsto \mathrm{Ric}(e_j) \wedge (e_j \lrcorner \omega)$ is positive too, hence taking the Hermitian product with ω in (20.11) and integrating over M yields

$$\int_M |\nabla\omega|^2 + H(\mathrm{Ric}(\omega), \omega)dv = 0,$$

showing that ω has to vanish. $\qquad\square$

20.3. Exercises

(1) Let A be a positive definite symmetric endomorphism of TM. Show that its extension to $\Lambda^p M \otimes \mathbb{C}$ defined by $\omega \mapsto A(e_j) \wedge (e_j \lrcorner \omega)$ is positive definite as well.

(2) Prove the following real version of the Weitzenböck formula:
$$\Delta\omega = \nabla^*\nabla\omega + \mathcal{R}\omega, \qquad \forall\, \omega \in \Omega^p M,$$
where $\mathcal{R}$ is the endomorphism of $\Omega^p M$ defined by
$$\mathcal{R}(\omega) := -e_j \wedge e_k \lrcorner (R(e_j, e_k)(\omega)).$$

(3) Applying the above identity to 1-forms, prove the *Bochner formula*
$$\Delta\omega = \nabla^*\nabla\omega + \mathrm{Ric}(\omega), \qquad \forall\, \omega \in \Omega^1 M.$$

(4) Prove that there are no global holomorphic forms on the complex projective space.

CHAPTER 21

The Hirzebruch–Riemann–Roch formula

21.1. Positive line bundles

In order to state another application of the Weitzenböck formula we have to make the following:

DEFINITION 21.1. *A real $(1,1)$-form φ on a complex manifold (M, g, J) is called* positive *(resp.* negative*) if the symmetric tensor $\varphi(\cdot, J\cdot)$ is positive (resp. negative) definite. A cohomology class in $H^{1,1}(M) \cap H^2(M, \mathbb{R})$ is called* positive *(resp.* negative*) if it can be represented by a positive (resp. negative) $(1,1)$-form. A holomorphic line bundle L over a compact complex manifold is called* positive *(resp.* negative*) if there exists a Hermitian structure on L with Chern connection ∇ and curvature form R^∇ such that iR^∇ is a positive (resp. negative) $(1,1)$-form.*

The positivity of a holomorphic line bundle is a cohomological property:

LEMMA 21.2. *A holomorphic line bundle L over a compact complex manifold M is positive if and only if its first Chern class is positive.*

PROOF. One direction is clear from the definition. Suppose, conversely, that $c_1(L)$ is positive. That means that there exists a positive $(1,1)$-form ω and a Hermitian structure h on L whose Chern connection ∇ has curvature R^∇ such that $[iR^\nabla] = [\omega]$ (the factor 2π can obviously be skipped). From the global $i\partial\bar\partial$-lemma, there exists a real function u such that $iR^\nabla = \omega + i\partial\bar\partial u$. We now use the formula (16.4) which gives the curvature of the Chern connection in terms of the square norm of an arbitrary local holomorphic section σ:

$$R^\nabla = -\partial\bar\partial \log h(\sigma, \sigma).$$

The curvature of the Chern connection $\tilde\nabla$ associated to $\tilde h := he^u$ thus satisfies for every local holomorphic section σ:

$$iR^{\tilde\nabla} = -i\partial\bar\partial(\log \tilde h(\sigma, \sigma)) = -i\partial\bar\partial(\log h(\sigma, \sigma)) - i\partial\bar\partial u = iR^\nabla - i\partial\bar\partial u = \omega,$$

showing that L is positive. $\qquad\qquad\square$

In order to get a feeling for this notion, notice that the Kähler form of a Kähler manifold is positive, as well as the Ricci form of a Kähler manifold with positive Ricci tensor. From Proposition 17.4 we know that the canonical bundle K of a Kähler manifold has curvature $i\rho$. In particular, if the Ricci tensor is positive definite, K is a negative line bundle.

THEOREM 21.3. *A negative holomorphic line bundle L over a compact Kähler manifold has no non-vanishing holomorphic section.*

PROOF. Consider a Hermitian structure on L with Chern connection ∇, such that iR^∇ is negative. Taking $p = 0$ and $E = L$ in (20.8) shows that every holomorphic section ξ of E satisfies

$$0 = 2\bar{\partial}^*\bar{\partial}\xi = \nabla^*\nabla\xi + \frac{i}{2}R^\nabla(Je_j, e_j)\xi. \tag{21.1}$$

By hypothesis we have $iR^\nabla(X, Y) = A(JX, Y)$, with A negative definite. Thus

$$\frac{i}{2}R^\nabla(Je_j, e_j) = -\frac{1}{2}A(e_j, e_j) = -\frac{1}{2}\mathrm{Tr}(A)$$

is a strictly positive function on M. Consequently, taking the Hermitian product with ξ in (21.1) and integrating over M shows that ξ has to vanish. $\square$

This result is consistent with our previous calculations on $\mathbb{C}P^m$. Since the Fubini–Study metric has positive Einstein constant, the canonical bundle of $\mathbb{C}P^m$ is negative. By Proposition 9.4, the tautological line bundle of $\mathbb{C}P^m$ is negative too. On the other hand, we already noticed (Exercise (3) in Chapter 9) that the tautological bundle has no holomorphic section.

21.2. The Hirzebruch–Riemann–Roch formula

Let $E \to M$ be a holomorphic vector bundle over some compact complex manifold M^{2m}. We denote as before by $\Omega^{0,k}(E) := \Gamma(\Lambda^{0,k}M \otimes E)$ the space of E-valued $(0, k)$-forms on M. Consider the following elliptic complex

$$\Omega^{0,0}(E) \xrightarrow{\bar{\partial}} \Omega^{0,1}(E) \xrightarrow{\bar{\partial}} \cdots \xrightarrow{\bar{\partial}} \Omega^{0,m}(E). \tag{21.2}$$

We define the cohomology groups

$$H^q(M, E) := \frac{\mathrm{Ker}(\bar{\partial} : \Omega^{0,q}(E) \to \Omega^{0,q+1}(E))}{\bar{\partial}(\Omega^{0,q-1}(E))}.$$

By analogy with the usual (untwisted) case, we denote

$$H^{p,q}(M, E) := H^q(M, \Lambda^{p,0}M \otimes E).$$

For every Hermitian structure on E and Hermitian metric on M, we consider the formal adjoint $\bar{\partial}^*$ of $\bar{\partial}$, and define the space of harmonic E-valued $(0, q)$-forms on M by

$$\mathcal{H}^q(E) := \{\omega \in \Omega^{0,q}(E) \mid \bar{\partial}\omega = 0, \ \bar{\partial}^*\omega = 0\}.$$

The analogue of the Dolbeault decomposition theorem holds true in this case and as a corollary we have:

THEOREM 21.4. *For every Hermitian bundle E on a compact Hermitian manifold M, the cohomology groups $H^q(M, E)$ are isomorphic with the spaces of harmonic E-valued $(0, q)$-forms:*

$$H^q(M, E) \simeq \mathcal{H}^q(E).$$

The elliptic complex (21.2) can be encoded in a single elliptic first order differential operator

$$\bar{\partial} + \bar{\partial}^* : \Omega^{0,\text{even}}(E) \to \Omega^{0,\text{odd}}(E).$$

The *index* of the elliptic complex (21.2) is defined to be the index of this elliptic operator:

$$\text{Ind}(\bar{\partial} + \bar{\partial}^*) := \dim(\text{Ker}(\bar{\partial} + \bar{\partial}^*)) - \dim(\text{Coker}(\bar{\partial} + \bar{\partial}^*)).$$

The holomorphic Euler characteristic $\Xi(M, E)$ is defined by

$$\Xi(M, E) := \sum_{k=0}^{m}(-1)^k \dim H^k(M, E)$$

and is nothing else but the above index. If E is the trivial line bundle, the holomorphic Euler characteristic $\Xi(M, E)$ is simply denoted by $\Xi(M) := \sum_{k=0}^{m}(-1)^k h^{0,k}(M)$.

THEOREM 21.5. (The Hirzebruch–Riemann–Roch formula) *Let E be a holomorphic bundle over a compact complex manifold M. The holomorphic Euler characteristic of E can be computed as follows*

$$\Xi(M, E) = \int_M \text{Td}(M)\text{ch}(E),$$

where $\text{Td}(M)$ is the Todd class of the tangent bundle of M and $\text{ch}(E)$ is the Chern character of E.

The Todd class and the Chern character are characteristic classes of the corresponding vector bundles that we will not define explicitly. The only thing that we will use in the sequel is that they satisfy the naturality axiom with respect to pull-backs.

For a proof of Theorem 21.5 and more details about characteristic classes see [5], Section 25.4.

We will give two applications of the Hirzebruch–Riemann–Roch formula, both concerning the fundamental group of Kähler manifolds under suitable positivity assumptions of the Ricci tensor. The first one is a theorem due to Kobayashi:

THEOREM 21.6. [9] *A compact Kähler manifold with positive definite Ricci tensor is simply connected.*

PROOF. Theorem 20.5 shows that there is no non-trivial holomorphic $(p, 0)$-form on M, so $h^{p,0}(M) = 0$ for $p > 0$. Of course, the holomorphic $(0, 0)$-forms are just the constant functions, so $h^{0,0}(M) = 1$. Since M is Kähler, we have $h^{p,0}(M) = h^{0,p}(M)$, which yields $\Xi(M) = 1$.

Now, by Myers' Theorem ([10], p. 88), the fundamental group of M is finite. Let $\tilde{M}$ be the universal cover of M, which is therefore compact too. Applying the previous argument to $\tilde{M}$ we get $\Xi(\tilde{M}) = 1$. But if $\pi : \tilde{M} \to M$ denotes the covering projection, we have, by naturality, $\mathrm{Td}(\tilde{M}) = \pi^* \mathrm{Td}(M)$, and an easy exercise shows that for every top degree form ω on M one has

$$\int_{\tilde{M}} \pi^* \omega = k \int_M \omega,$$

where k denotes the number of sheets of the covering. This shows that $k = 1$, so M is simply connected. □

COROLLARY 21.7. *If the first Chern class of a compact Kähler manifold is positive, then M is simply connected.*

PROOF. By the Calabi–Yau theorem M has a Kähler metric with positive Ricci curvature, so the result follows from Theorem 21.6. □

Our second application concerns Ricci-flat Kähler manifolds. By Theorem 17.5, a compact Kähler manifold M^{2m} is Ricci-flat if and only if the restricted holonomy group $\mathrm{Hol}_0(M)$ is a subgroup of SU_m. A compact Kähler manifold M with $\mathrm{Hol}_0(M) = \mathrm{SU}_m$ is called a *Calabi–Yau manifold*.

THEOREM 21.8. *Let M^{2m} be a Calabi–Yau manifold. If m is odd, then $\mathrm{Hol}(M) = \mathrm{SU}_m$, so there exists a global holomorphic $(m, 0)$-form even if M is not simply connected. If m is even, then either M is simply connected, or $\pi_1(M) = \mathbb{Z}_2$ and M carries no global holomorphic $(m, 0)$-form.*

PROOF. Let $\tilde{M}$ be the universal covering of M. Since $\tilde{M}$ has irreducible holonomy, the Cheeger–Gromoll theorem (cf. [1], p. 168) shows that it is compact. By Theorem 20.5, every holomorphic form on M is parallel, and thus corresponds to a fixed point of the holonomy representation. It is easy to check that SU_m has only two invariant one-dimensional complex subspaces on $(p, 0)$-forms, one for $p = 0$ and one for $p = m$. Thus

$$\Xi(\tilde{M}) = \begin{cases} 0, & \text{for } m \text{ odd}, \\ 2, & \text{for } m \text{ even}. \end{cases}$$

Moreover, $\Xi(\tilde{M}) = k\Xi(M)$, where k is the order of the fundamental group of M. This shows that $\Xi(M) = 0$ for m odd. Since $h^{p,0}(M) = 0$ for $0 < p < m$, we necessarily have $h^{m,0}M = 1$, so M carries a global holomorphic $(m, 0)$-form.

If m is even, then either M is simply connected, or $k = 2$ and $\Xi(M) = 1$. In this last case, we necessarily have $h^{m,0}M = 0$, so M carries no global holomorphic $(m, 0)$-form. $\square$

21.3. Exercises

(1) Prove the *Kodaira–Serre duality*:
$$H^q(M, E) \simeq H^{m-q}(M, E^* \otimes K_M)$$
for every holomorphic vector bundle E over a compact complex manifold M. *Hint:* Choose a Hermitian structure on E and a Hermitian metric on M and use Theorem 21.4.

(2) Prove that the operator
$$\bar\partial + \bar\partial^* : \Omega^{0,\text{even}}(E) \to \Omega^{0,\text{odd}}(E)$$
is elliptic, in the sense that its principal symbol applied to any non-zero real 1-form is an isomorphism.

(3) Prove that the index of the above defined operator is equal to the holomorphic Euler characteristic $\Xi(M, E)$.

(4) Let $\pi : \tilde M \to M$ be a k-sheet covering projection between compact oriented manifolds. Prove that for every top degree form ω on M one has
$$\int_{\tilde M} \pi^*\omega = k \int_M \omega.$$
Hint: Start by showing that to any open cover $\{U_i\}$ of M one can associate a closed cover $\{C_j\}$ such that for every j there exists some i with $C_j \subset U_i$ and such that the interiors of C_j and C_k are disjoint for every $j \neq k$.

(5) Show that the representation of SU_m on $\Lambda^p \mathbb{C}^m$ given by
$$A(v_1 \wedge \cdots \wedge v_p) := Av_1 \wedge \cdots \wedge Av_p,$$
has no invariant one-dimensional subspace for $1 \leq p \leq m - 1$.

CHAPTER 22

Further vanishing results

22.1. The Lichnerowicz formula for Kähler manifolds

Let L be a holomorphic Hermitian line bundle over some Kähler manifold (M^{2m}, g, J) with scalar curvature S. We consider the curvature term in the Weitzenböck formula on sections of $\Lambda^{0,k} M \otimes L$, and claim that this term becomes very simple in the case where L is a square root of the canonical bundle. The reader familiar with spin geometry will notice that in this case $\Lambda^{0,*} M \otimes K^{1/2}$ is the *spin bundle* of M and the operator $\sqrt{2}(\partial + \bar{\partial})$ is exactly the Dirac operator.

Let us denote by $i\alpha$ the curvature of the Chern connection of L. We express (20.7) as $\mathcal{R} = \mathcal{R}_1 + \mathcal{R}_2$. The first term, $\mathcal{R}_1$, applied to some section $\omega \otimes \xi \in \Omega^{0,k} M \otimes E$, can be computed as follows

$$\mathcal{R}_1(\omega \otimes \xi) \ := \ \frac{i}{2} \tilde{R}(Je_j, e_j)(\omega \otimes \xi) = \frac{i}{2}\left(2(\rho^{(k)}\omega) \otimes \xi + i\alpha(Je_j, e_j)\omega \otimes \xi \right)$$

$$= \ i(\rho^{(k)}\omega) \otimes \xi - \frac{1}{2}\alpha(Je_j, e_j)\omega \otimes \xi.$$

In order to compute the second curvature term we make use of the following algebraic result:

LEMMA 22.1. *The Riemannian curvature operator satisfies*

$$(e_j - iJe_j) \wedge (e_k + iJe_k) \lrcorner R(e_j, e_k)\omega = 4i\rho^{(k)}\omega$$

for every $(0, k)$-form ω.

PROOF. Since the interior product of a $(1, 0)$-vector and a $(0, k)$-form vanishes we obtain

$$X \lrcorner \omega = iJX \lrcorner \omega, \qquad \forall\, \omega \in \Omega^{0,k} M, \ X \in TM. \tag{22.1}$$

The forms $R(e_j, e_k)\omega$ are still $(0, k)$-forms, since the connection preserves the type decomposition of forms. By changing e_j to Je_j and then e_k to Je_k we get

$$
\begin{aligned}
e_j \wedge (e_k + iJe_k) \lrcorner R(e_j, e_k)\omega &= Je_j \wedge (e_k + iJe_k) \lrcorner R(Je_j, e_k)\omega \\
&= -Je_j \wedge (e_k + iJe_k) \lrcorner R(e_j, Je_k)\omega \\
&= -iJe_j \wedge (e_k + iJe_k) \lrcorner R(e_j, e_k)\omega.
\end{aligned}
$$

Thus

$$(e_j - iJe_j) \wedge (e_k + iJe_k) \lrcorner R(e_j, e_k)\omega = 2e_j \wedge (e_k + iJe_k) \lrcorner R(e_j, e_k)\omega$$
$$= 4e_j \wedge e_k \lrcorner R(e_j, e_k)\omega.$$

Now, using (22.1) twice yields

$$R(e_j, e_k, e_l, e_s)e_j \wedge e_k \wedge e_s \lrcorner e_l \lrcorner \omega = -R(e_j, e_k, Je_l, Je_s)e_j \wedge e_k \wedge e_s \lrcorner e_l \lrcorner \omega$$
$$= -R(e_j, e_k, e_l, e_s)e_j \wedge e_k \wedge e_s \lrcorner e_l \lrcorner \omega,$$

so this expression vanishes. From the first Bianchi identity we then obtain

$$R(e_j, e_k, e_l, e_s)e_j \wedge e_s \wedge e_k \lrcorner e_l \lrcorner \omega = R(e_j, e_l, e_k, e_s)e_j \wedge e_s \wedge e_k \lrcorner e_l \lrcorner \omega$$
$$+ R(e_j, e_s, e_l, e_k)e_j \wedge e_s \wedge e_k \lrcorner e_l \lrcorner \omega$$
$$= -R(e_j, e_l, e_k, e_s)e_j \wedge e_s \wedge e_l \lrcorner e_k \lrcorner \omega$$

whence

$$R(e_j, e_l, e_k, e_s)e_j \wedge e_s \wedge e_k \lrcorner e_l \lrcorner \omega = 0.$$

Finally we get

$$(e_j - iJe_j) \wedge (e_k + iJe_k) \lrcorner R(e_j, e_k)\omega = 4e_j \wedge e_k \lrcorner R(e_j, e_k)\omega$$
$$= 4R(e_j, e_k, e_l, e_s)e_j \wedge e_k \lrcorner (e_s \wedge e_l \lrcorner \omega)$$
$$= -4\mathrm{Ric}(e_j, e_l)e_j \wedge e_l \lrcorner \omega = 4i\mathrm{Ric}(e_j, Je_l)e_j \wedge e_l \lrcorner \omega = 4i\rho^{(k)}\omega.$$

$\square$

For every $(1,1)$-form α and $(0,k)$-form ω we have as before

$$\alpha(e_j, e_k)(e_j - iJe_j) \wedge (e_k + iJe_k) \lrcorner \omega = 2\alpha(e_j, e_k)e_j \wedge (e_k + iJe_k) \lrcorner \omega$$
$$= 4\alpha(e_j, e_k)e_j \wedge e_k \lrcorner \omega = -4\alpha^{(k)}(\omega).$$

The second term in (20.7) thus reads

$$\mathcal{R}_2(\omega \otimes \xi) := -\frac{1}{2}(e_j - iJe_j) \wedge (e_k + iJe_k) \lrcorner (\tilde{R}(e_j, e_k)(\omega \otimes \xi))$$
$$= -\frac{1}{2}(e_j - iJe_j) \wedge (e_k + iJe_k) \lrcorner ((R(e_j, e_k)\omega) \otimes \xi$$
$$+ i\alpha(e_j, e_k)\omega \otimes \xi)$$
$$= -2i\rho^{(k)}(\omega) \otimes \xi + 2i\alpha^{(k)}(\omega) \otimes \xi.$$

Suppose that the curvature of the line bundle L satisfies

$$R^L := i\alpha = \frac{1}{2}i\rho.$$

The formulas above show that the curvature term in the Weitzenböck formula on $\Omega^{0,k}(L)$ satisfies

$$
\begin{aligned}
\mathcal{R}(\omega \otimes \xi) &= (\mathcal{R}_1 + \mathcal{R}_2)(\omega \otimes \xi) = i\rho^{(k)}(\omega) \otimes \xi - \frac{1}{2}\alpha(Je_j, e_j)\omega \otimes \xi \\
&\quad -2i\rho^{(k)}(\omega) \otimes \xi + 2i\alpha^{(k)}(\omega) \otimes \xi \\
&= -\frac{1}{4}\rho(Je_j, e_j)\omega \otimes \xi = \frac{S}{4}\omega \otimes \xi,
\end{aligned}
$$

where $S = \mathrm{Tr}(\mathrm{Ric})$ denotes the scalar curvature. This proves:

THEOREM 22.2. (Lichnerowicz formula) *Let $L = K^{1/2}$ be a square root of the canonical bundle of a Kähler manifold (M^{2m}, h, J), in the sense that L has a Hermitian structure H such that (K, h) is isomorphic to $L \otimes L$ with the induced tensor product Hermitian structure. Then, if Ψ is a section of the bundle*

$$
\Sigma M := (\Lambda^{0,0}M \oplus \cdots \oplus \Lambda^{0,m}M) \otimes L
$$

and $D := \sqrt{2}(\bar{\partial} + \bar{\partial}^*)$ is the Dirac operator on ΣM, the following formula holds

$$
D^2\Psi = \nabla^*\nabla\Psi + \frac{S}{4}\Psi.
$$

The Lichnerowicz formula is valid in a more general setting (on all spin manifolds, not necessarily Kähler), and it has important applications in geometry and topology (see [4], [14]).

22.2. The Kodaira vanishing theorem

Let L be a positive holomorphic line bundle over a compact complex manifold (M^{2m}, J). By definition, L carries a Hermitian structure whose Chern connection ∇ has curvature R^∇ with $iR^\nabla > 0$. Consider the Kähler metric h on M whose Kähler form is just iR^∇. By a slight abuse of language, we denote by $\partial : \Omega^{p,q}(L) \to \Omega^{p+1,q}M(L)$ the extension of $\nabla^{1,0}$ to L-valued forms. Notice that, whilst $\bar{\partial}$ is an intrinsic operator, ∂ depends of course on the Hermitian structure on L. We apply the Weitzenböck formula to some section $\omega \otimes \xi$ of $\Lambda^{p,q}M \otimes L$:

$$
2(\bar{\partial}^*\bar{\partial} + \bar{\partial}\bar{\partial}^*)(\omega \otimes \xi) = \nabla^*\nabla(\omega \otimes \xi) + \mathcal{R}(\omega \otimes \xi). \tag{22.2}
$$

The same computation actually yields the dual formula

$$
2(\partial^*\partial + \partial\partial^*)(\omega \otimes \xi) = \nabla^*\nabla(\omega \otimes \xi) + \bar{\mathcal{R}}(\omega \otimes \xi), \tag{22.3}
$$

where $\bar{\mathcal{R}}$ is the complex conjugate of $\mathcal{R}$. Subtracting these two equations yields

$$
2(\bar{\partial}^*\bar{\partial} + \bar{\partial}\bar{\partial}^*)(\omega \otimes \xi) = 2(\partial^*\partial + \partial\partial^*)(\omega \otimes \xi) + (\mathcal{R} - \bar{\mathcal{R}})(\omega \otimes \xi). \tag{22.4}
$$

We now compute this curvature term using (20.7) and a local h-orthonormal frame $\{e_i\}$ parallel at some point:

$$(\mathcal{R} - \tilde{\mathcal{R}})(w \otimes \xi)$$

$$= i\tilde{R}(Je_j, e_j)(w \otimes \xi) - \frac{1}{2}(e_j - iJe_j) \wedge (e_k + iJe_k) \lrcorner \tilde{R}(e_j, e_k)(w \otimes \xi)$$

$$+ \frac{1}{2}(e_j + iJe_j) \wedge (e_k - iJe_k) \lrcorner \tilde{R}(e_j, e_k)(w \otimes \xi)$$

$$= i\tilde{R}(Je_j, e_j)(w \otimes \xi) + iJe_j \wedge e_k \lrcorner \tilde{R}(e_j, e_k)(w \otimes \xi)$$

$$-ie_j \wedge Je_k \lrcorner \tilde{R}(e_j, e_k)(w \otimes \xi)$$

$$= i\tilde{R}(Je_j, e_j)(w \otimes \xi) + 2iJe_j \wedge e_k \lrcorner \tilde{R}(e_j, e_k)(w \otimes \xi)$$

$$= 2i\rho(w) \otimes \xi + iw \otimes R^{\nabla}(Je_j, e_j)\xi + 2iJe_j \wedge e_k \lrcorner R(e_j, e_k)w \otimes \xi$$

$$+ 2iJe_j \wedge e_k \lrcorner w \otimes R^{\nabla}(e_j, e_k)\xi$$

$$= 2i\rho(w) \otimes \xi - 2mw \otimes \xi + 2iJe_j \wedge e_k \lrcorner R(e_j, e_k)w \otimes \xi + 2(p+q)w \otimes \xi,$$

where the last equality uses the fact that $iR^{\nabla}(\cdot, \cdot) = h(J\cdot, \cdot)$. On the other hand, the expression $Je_j \wedge e_k \lrcorner R(e_j, e_k)w$ can be computed as follows:

$$Je_j \wedge e_k \lrcorner R(e_j, e_k)w$$

$$= Je_j \wedge e_k \lrcorner R(e_j, e_k)e_l \wedge e_l \lrcorner w$$

$$= -\text{Ric}(e_j, e_l)Je_j \wedge e_l \lrcorner w - Je_j \wedge R(e_j, e_k)e_l \wedge e_k \lrcorner e_l \lrcorner w$$

$$= -\rho(w) - R(e_j, e_k, e_l, e_s)Je_j \wedge e_s \wedge e_k \lrcorner e_l \lrcorner w.$$

From the first Bianchi identity we get

$$2R(e_j, e_k, e_l, e_s) \; Je_j \wedge e_s \wedge e_k \lrcorner e_l \lrcorner w$$

$$= R(e_j, e_k, e_l, e_s)Je_j \wedge e_s \wedge e_k \lrcorner e_l \lrcorner w$$

$$+ R(e_j, e_l, e_k, e_s)Je_j \wedge e_s \wedge e_l \lrcorner e_k \lrcorner w$$

$$= R(e_l, e_k, e_j, e_s)Je_j \wedge e_s \wedge e_k \lrcorner e_l \lrcorner w$$

$$= -R(e_l, e_k, Je_j, e_s)e_j \wedge e_s \wedge e_k \lrcorner e_l \lrcorner w = 0,$$

where the last expression vanishes because $R(\cdot, \cdot, J\cdot, \cdot)$ is symmetric in the last two arguments.

This shows that $Je_j \wedge e_k \lrcorner R(e_j, e_k)w = -\rho(w)$, so the previous calculation yields

$$2(\bar{\partial}^*\bar{\partial} + \bar{\partial}\bar{\partial}^*)(w \otimes \xi) = 2(\partial^*\partial + \partial\partial^*)(w \otimes \xi) + 2(p + q - m)(w \otimes \xi).$$

After taking the Hermitian product with $w \otimes \xi$ (which we denote by σ for simplicity) and integrating over M we get

$$\int_M |\bar{\partial}\sigma|^2 + |\bar{\partial}^*\sigma|^2 dv = \int_M |\partial\sigma|^2 + |\partial^*\sigma|^2 + (p + q - m)|\sigma|^2 dv. \qquad (22.5)$$

If σ is a harmonic L-valued form, the left hand side term in (22.5) vanishes, thus proving:

THEOREM 22.3. (Kodaira vanishing theorem) *If L is a positive holomorphic line bundle on a compact Kähler manifold M, one has $H^{p,q}(M, L) = 0$ whenever $p + q > m$.*

22.3. Exercises

(1) Let L be a negative holomorphic line bundle on a compact complex manifold M of complex dimension m. Prove that L admits no holomorphic sections. More generally, prove that $H^q(M, L) = 0$ for $q < m$. *Hint:* Apply the Kodaira–Serre duality.

(2) Let L be a positive holomorphic line bundle on a compact complex manifold M of complex dimension m. Prove that there exists a positive integer $k(L) \in \mathbb{N}$ such that $H^p(M, L^k) = 0$ for all $p > 0$ and $k \geq k(L)$.

CHAPTER 23

Ricci-flat Kähler metrics

23.1. Hyperkähler manifolds

The aim of this chapter is to obtain the classification (up to finite coverings) of compact Ricci-flat Kähler manifolds. We start with the following:

DEFINITION 23.1. *A Riemannian manifold* (M^n, g) *is called* hyperkähler *if there exist three complex structures* I, J, K *on* M *satisfying* $K = IJ$ *such that* g *is a Kähler metric with respect to each of these complex structures.*

It is clear that a metric is hyperkähler if and only if it is Kähler with respect to two anti-commuting complex structures. In the irreducible case, this can be weakened as follows:

PROPOSITION 23.2. *Let* (M^n, g, ∇) *be a locally irreducible Riemannian manifold. If* g *is Kähler with respect to two complex structures* J *and* J_1, *and if* J_1 *is different from* J *and* $-J$, *then* (M, g) *is hyperkähler.*

PROOF. The endomorphism $JJ_1 + J_1J$ is symmetric and ∇-parallel on M, so by the irreducibility hypothesis, it has to be constant:

$$JJ_1 + J_1J = \alpha \mathrm{Id}_{TM}, \qquad \alpha \in \mathbb{R}. \tag{23.1}$$

From the Cauchy–Schwartz inequality we get

$$\alpha^2 = |\alpha \mathrm{Id}|^2 = |JJ_1 + J_1J|^2 \le 2(|JJ_1|^2 + |J_1J|^2) \le 4|J|^2|J_1|^2 = 4,$$

where the norm considered here is the operator norm. The equality case can only hold if $JJ_1 = \beta J_1 J$ for some real number β. Together with (23.1) this shows that $JJ_1 = \gamma \mathrm{Id}$ for some real number γ, so $J_1 = \pm J$, which was excluded in the hypothesis. Therefore we have $\alpha^2 < 4$. We then compute using (23.1)

$$(J_1 + JJ_1J)^2 = (\alpha^2 - 4)\mathrm{Id}_{TM},$$

so the ∇-parallel skew-symmetric endomorphism

$$I := \frac{1}{\sqrt{4 - \alpha^2}}(J_1 + JJ_1J)$$

defines a complex structure anti-commuting with J, with respect to which g is Kähler. □

Let $\mathbb{H}$ denote the field of quaternions and consider the identification of $\mathbb{C}^{2k}$ with $\mathbb{H}^k$ given by $(z_1, z_2) \mapsto z_1 + jz_2$. We denote by I, J and K the right product on $\mathbb{H}^k$ by i, j and k respectively, which correspond to the following endomorphisms of $\mathbb{C}^{2k}$:

$$I(z_1, z_2) := (iz_1, iz_2), \qquad J(z_1, z_2) := (-\bar{z}_2, \bar{z}_1), \qquad K(z_1, z_2) := (-i\bar{z}_2, i\bar{z}_1).$$

Let Sp_k denote the group of unitary transformations of $\mathbb{C}^{2k}$ (that is, preserving the canonical Hermitian product and commuting with I), which commute with J (and thus also with K). An easy computation using block matrices shows that

$$\mathrm{Sp}_k = \left\{ M = \begin{pmatrix} A & B \\ -\bar{B} & \bar{A} \end{pmatrix} \in \mathcal{M}_{2k}(\mathbb{C}) \, \middle| \, M\bar{M}^t = I_{2k} \right\}.$$

It is tautological that a $4k$-dimensional manifold is hyperkähler if and only if the bundle of orthonormal frames has a reduction to Sp_k.

LEMMA 23.3. $\mathrm{Sp}_k \subset \mathrm{SU}_{2k}$.

PROOF. By definition we have $\mathrm{Sp}_k \subset \mathrm{U}_{2k}$, so every matrix in Sp_k is diagonalizable as a complex matrix and its eigenvalues are complex numbers of unit norm. If v is an eigenvector of some $M \in \mathrm{Sp}_k$ with eigenvalue $\lambda \in S^1$ then Jv is an eigenvector with eigenvalue λ^{-1}:

$$MJv = JMv = J\lambda v = \bar{\lambda}Jv = \lambda^{-1}Jv.$$

The determinant of M (as a matrix in $\mathcal{M}_{2k}(\mathbb{C})$) is thus equal to 1. $\square$

This shows that every hyperkähler manifold is Ricci-flat. A hyperkähler manifold is called *strict* if it is locally irreducible.

Let now M be an arbitrary compact Ricci-flat Kähler manifold. By the Cheeger–Gromoll theorem ([1], p. 168), M is isomorphic to a quotient

$$M \simeq (M_0 \times \mathbb{T}^l)/\Gamma,$$

where M_0 is a compact simply connected Kähler manifold, $\mathbb{T}^l$ is a complex torus and Γ is a finite group of holomorphic transformations. Let $M_0 = M_1 \times \cdots \times M_s$ be the *de Rham decomposition* of M_0 (cf. [10], p. 187). Then M_j are compact Ricci-flat simply connected Kähler manifolds with irreducible holonomy for all j. By the Berger holonomy theorem ([1], p. 300) each M_j is either symmetric or Calabi–Yau or strict hyperkähler. A symmetric space which is Ricci-flat is automatically flat ([11], p. 231), so the M_j's are not symmetric. We thus have the following:

THEOREM 23.4. *A compact Ricci-flat Kähler manifold M is isomorphic to the quotient*

$$M \simeq (M_1 \times \cdots \times M_s \times M_{s+1} \times \cdots \times M_r \times \mathbb{T}^l)/\Gamma,$$

where M_j are simply connected compact Calabi–Yau manifolds for $j \leq s$, simply connected compact strict hyperkähler manifolds for $s + 1 \leq j \leq r$ and Γ is a finite group of holomorphic transformations.

23.2. Projective manifolds

A compact complex manifold (M^{2m}, J) is called *projective* if it can be holomorphically embedded in some complex projective space $\mathbb{CP}^N$. A well-known result of Chow states that a projective manifold is *algebraic*, that is, defined by a finite set of homogeneous polynomials in the complex projective space.

PROPOSITION 23.5. *Every projective manifold has a positive holomorphic line bundle.*

PROOF. Let Ω be the Kähler form of the Fubini–Study metric on $\mathbb{CP}^N$. It is easy to check (e.g. using (13.1)) that the hyperplane line bundle H on $\mathbb{CP}^N$ has a connection with curvature $-i\Omega$. The pull-back of this line bundle to any complex submanifold of $\mathbb{CP}^N$ is thus positive. $\square$

Conversely, we have the celebrated:

THEOREM 23.6. (Kodaira embedding theorem) *A compact complex manifold M with a positive holomorphic line bundle L is projective.*

A proof can be found in [3], p. 176. The main idea is to show that a suitable positive power L^k of L has a basis of holomorphic sections $\{\sigma_0, \ldots, \sigma_N\}$ such that the holomorphic mapping

$$M \to \mathbb{CP}^N \qquad x \mapsto [\sigma_0(x) : \ldots : \sigma_N(x)]$$

is a well-defined embedding.

COROLLARY 23.7. *Every Calabi–Yau manifold of complex dimension $m \geq 3$ is projective.*

PROOF. For every compact manifold M, let $\mathcal{A}$ and $\mathcal{A}^*$ be the sheaves ([3], p. 35) of smooth functions on M with values in $\mathbb{C}$ and $\mathbb{C}^*$ respectively. The exact sequence of sheaves

$$0 \to \mathbb{Z} \xrightarrow{2\pi i} \mathcal{A} \xrightarrow{\exp} \mathcal{A}^* \to 0$$

induces an exact sequence in Čech cohomology

$$\cdots \to H^1(M, \mathcal{A}) \to H^1(M, \mathcal{A}^*) \xrightarrow{c_1} H^2(M, \mathbb{Z}) \to H^2(M, \mathcal{A}) \to \cdots$$

The sheaf $\mathcal{A}$ is *fine* (that is, it admits a partition of unity), so $H^1(M, \mathcal{A}) = 0$ and $H^2(M, \mathcal{A}) = 0$, thus proving that c_1 is an isomorphism. Notice that $H^1(M, \mathcal{A}^*)$ is just the set of equivalence classes of complex line bundles over M, and the isomorphism above is given by the first Chern class ([3], p. 141).

This shows that for every integer cohomology class $\gamma \in H^2(M, \mathbb{Z})$, there exists a complex line bundle L with $c_1(L) = \gamma$. Moreover, if ω is any complex 2-form representing γ in real cohomology, there exists a connection ∇ on L such that $(i/2\pi)R^\nabla = \omega$. To see this, take any connection $\tilde{\nabla}$ on L with curvature $R^{\tilde{\nabla}}$. Then since $[\omega] = c_1(L)$ we get $[2\pi\omega] = [iR^{\tilde{\nabla}}]$, so there exists some 1-form θ such that $2\pi\omega = i(R^{\tilde{\nabla}} + d\theta)$. Clearly the curvature of $\nabla :=$ $\tilde{\nabla} + i\theta$ satisfies the desired equation. If the form ω is real and of type $(1,1)$, then the complex bundle L has a holomorphic structure, given by the $(0,1)$-part of the connection whose curvature is ω.

Let now M^{2m} be a Calabi–Yau manifold, $m > 2$. Since SU_m has no fixed point on $\Lambda^{2,0}\mathbb{C}^m$ (see the last exercise of Chapter 21), we deduce that there are no parallel $(2,0)$-forms on M, so by Theorem 20.5 we get $h^{2,0}(M) = 0$. By the Dolbeault decomposition theorem we obtain that any harmonic 2-form on M is of type $(1,1)$. Consider the Kähler form Ω of M. Since $H^2(M, \mathbb{Q})$ is dense in $H^2(M, \mathbb{R})$, and the space of positive harmonic $(1,1)$-forms is open in $\mathcal{H}^{1,1}(M, \mathbb{R}) = \mathcal{H}^2(M, \mathbb{R})$, we can approximate Ω by a positive harmonic $(1,1)$-form ω such that $[\omega] \in H^2(M, \mathbb{Q})$. By multiplying with the common denominator, we may suppose that $[\omega] \in H^2(M, \mathbb{Z})$. Then the argument above shows that there exists a holomorphic line bundle whose first Chern class is ω, thus a positive holomorphic line bundle on M. By the Kodaira embedding theorem, M is then projective. □

23.3. Exercises

(1) Show that every holomorphic vector field on a compact Ricci-flat Kähler manifold is parallel with respect to the Levi–Civita connection. Deduce that Calabi–Yau manifolds carry no holomorphic vector fields.

(2) Using Theorem 20.5 show that every holomorphic form on a compact Ricci-flat Kähler manifold (M^{2m}, h, J) is parallel. Deduce that up to constant multiples, the only holomorphic forms on a Calabi–Yau manifold are $1 \in \Omega^{0,0}M$ and the complex volume form in $\Omega^{m,0}M$.

(3) Let (M, J) be a complex manifold of even complex dimension $m = 2k$. A holomorphic $(2,0)$-form τ is called holomorphic symplectic if τ^k is nowhere zero on M. Show that if M is compact and admits a holomorphic symplectic form then $c_1(M, J) = 0$. Hint: The maximal exterior power of τ is a non-zero holomorphic section of the canonical bundle, which is thus trivial.

(4) Let (M, g, J) be a compact Kähler manifold carrying a holomorphic symplectic form τ. Show that there exists a Ricci-flat metric h on

M such that (M, h, J) is Kähler and that τ is parallel with respect to the Levi–Civita connection ∇ of h.

(5) With the notations of the previous exercise, show that if (M, h, J) is irreducible then τ defines a hyperkähler structure. *Hint:* Write $\tau = \alpha + i\beta$, where α and β are real ∇-parallel 2-forms, identified by h with skew-symmetric parallel endomorphisms. Show that $\beta = \alpha \circ J$, and use this to deduce that α and β anti-commute with J. Conclude by an argument similar to Proposition 23.2.

CHAPTER 24

Explicit examples of Calabi–Yau manifolds

24.1. Divisors

Let M be a compact complex manifold. An *analytic hypersurface* of M is a subset $V \in M$ such that for every $x \in V$ there exists an open set $U_x \subset M$ containing x and a holomorphic function f_x defined on U_x such that $V \cap U_x$ is the zero-set of f_x. Such an f_x is called a *local defining function* for V near x. The quotient of any two local defining functions around x is a non-vanishing holomorphic function around x.

An analytic hypersurface V is called *irreducible* if it can not be written as the union of two smaller analytic hypersurfaces. Every analytic hypersurface is a finite union of its irreducible components.

If V is an irreducible analytic hypersurface, with local defining function φ_x around some $x \in V$, then for every holomorphic function f around x, the *order* of f along V at x is defined to be the largest positive integer a such that f/φ_x^a is holomorphic around x. It can be shown that the order of f is a well-defined positive integer, which does not depend on x, and is denoted by $o(f, V)$ (see [3], p. 130).

DEFINITION 24.1. *A divisor D in a compact complex manifold M is a finite formal sum with integer coefficients of irreducible analytic hypersurfaces of M.*

$$D := \sum_i a_i V_i, \qquad a_i \in \mathbb{Z}.$$

A divisor D is called effective *if $a_i \geq 0$ for all i.*

The set of divisors has clearly the structure of a commutative group.

A *meromorphic function* on a complex manifold M is an equivalence class of collections $(U_\alpha, f_\alpha, g_\alpha)_{\alpha \in I}$ where $\{U_\alpha\}$ is an open covering of M, and f_α, g_α are holomorphic functions defined on U_α such that $f_\alpha g_\beta = f_\beta g_\alpha$ on $U_\alpha \cap U_\beta$ for all $\alpha, \beta \in I$. Two such collections $(U_\alpha, f_\alpha, g_\alpha)_{\alpha \in I}$ and $(U'_\beta, f'_\beta, g'_\beta)_{\beta \in J}$ are equivalent if $f_\alpha g'_\beta = f'_\beta g_\alpha$ on $U_\alpha \cap U'_\beta$ for all $\alpha \in I$, $\beta \in J$. A meromorphic function can be viewed around any $x \in U_a$ as f_α / g_α, provided that g_a does not vanish at x.

EXAMPLE. The expression "z_1/z_2" is in fact a meromorphic function on $\mathbb{CP}^2$, defined by the collection

$$(U_0, \frac{z_1}{z_0}, \frac{z_2}{z_0}), \quad (U_1, 1, \frac{z_2}{z_1}), \quad (U_2, \frac{z_1}{z_2}, 1),$$

where U_i denote as usual the canonical holomorphic atlas of $\mathbb{CP}^2$. This example illustrates the fact that a meromorphic function on M *does not* induce a well-defined holomorphic function $M \to \mathbb{CP}^1 = \mathbb{C} \cup \{\infty\}$.

If L is a holomorphic line bundle over M, we define similarly a *meromorphic section* of L as an equivalence class of collections $(U_\alpha, \sigma_\alpha, g_\alpha)_{\alpha \in I}$ where σ_α is a local holomorphic section of L over U_α and g_α is a holomorphic function on U_α, such that $\sigma_\alpha g_\beta = \sigma_\beta g_\alpha$ on $U_\alpha \cap U_\beta$ for all $\alpha, \beta \in I$.

A meromorphic function h defines a divisor (h) in a canonical way by

$$(h) := (h)_0 - (h)_\infty,$$

where $(h)_0$ and $(h)_\infty$ denote the zero-locus (resp. the pole-locus) of h taken with multiplicities.

More precisely, for every x in M, one can write the function h as $h = f_x/g_x$ near x. For each irreducible analytic hypersurface V containing x, we define the order of h along V at x to be $o(f_x, V) - o(g_x, V)$, which turns out to be independent of x, and is denoted by $o(h, V)$. Then

$$(h) = \sum_V o(h, V)V,$$

where the above sum is finite since for every open set U_x where $h = f_x/g_x$, there are only finitely many irreducible analytic hypersurfaces along which f_x of g_x have non-vanishing order.

Similarly, if σ is a global meromorphic section of a line bundle L, one can define the order $o(\sigma, V)$ of σ along any irreducible analytic hypersurface V using local trivializations of L. This clearly does not depend on the chosen trivialization, since the transition maps do not vanish, so they do not contribute to the order. As before, one defines a divisor (σ) on M by

$$(\sigma) = \sum_V o(\sigma, V)V.$$

If $D = \sum a_i V_i$ and f_i are local defining functions for V_i near some $x \in M$ (of course we can take $f_i = 1$ if V_i does not contain x), then the meromorphic function

$$\prod f_i^{a_i}$$

is called a *local defining function* for D around x.

DEFINITION 24.2. *Two divisors D and D' are called* linearly equivalent *if there exists some meromorphic function h such that*

$$D = D' + (h).$$

In this case we write $D \equiv D'$.

Clearly two meromorphic sections σ and σ' of L define linearly equivalent divisors $(\sigma) = (\sigma') + (h)$, where h is the meromorphic function defined by $\sigma = \sigma' h$.

24.2. Line bundles and divisors

To any divisor D we will associate a holomorphic line bundle $[D]$ on M in the following way. Take an open covering U_α of M and local defining meromorphic functions h_α for D defined on U_α. We define $[D]$ to be the holomorphic line bundle on M with transition functions $g_{\alpha\beta} := h_\alpha/h_\beta$. It is easy to check that $g_{\alpha\beta}$ are non-vanishing holomorphic functions on $U_\alpha \cap U_\beta$ satisfying the cocycle conditions, and that the equivalence class of $[D]$ does not depend on the choice of h_α.

EXAMPLE. Let H denote the hyperplane $\{z_0 = 0\}$ in $\mathbb{C}P^m$ and consider the usual open covering $U_\alpha = \{z_\alpha \neq 0\}$ of $\mathbb{C}P^m$. Then 1 is a local defining function for H on U_0 and z_0/z_α are local defining functions on U_α. The line bundle $[H]$ has thus transition functions $g_{\alpha\beta} = z_\beta/z_\alpha$, which are exactly the transition function of the *hyperplane line bundle* introduced in Exercise (4), Chapter 9. This justifies its denomination.

If D and D' are divisors, then clearly $[-D] = [D]^{-1}$ and $[D + D'] = [D] \otimes [D']$. We call $\mathrm{Div}(M)$ the group of divisors on M, and $\mathrm{Pic}(M) := H^1(M, \mathcal{O})$ the *Picard group* of equivalence classes of holomorphic line bundles (where $\mathcal{O}$ denotes the sheaf of holomorphic functions). Then the arguments above show that there exists a group homomorphism

$$[\,] : \mathrm{Div}(M) \to \mathrm{Pic}(M) \qquad D \mapsto [D].$$

Notice that the line bundle associated to a divisor (h) is trivial for every meromorphic function h. Indeed, for any open cover U_α on M, $h|_{U_\alpha}$ is a local defining function for the divisor (h) on U_α, so the transition functions for the line bundle $[(h)]$ are equal to 1 on any intersection $U_\alpha \cap U_\beta$. Thus $[\,]$ descends to a group homomorphism

$$[\,] : (\mathrm{Div}(M)/_\equiv) \longrightarrow \mathrm{Pic}(M).$$

Suppose now that $[D] = 0$ for some divisor D on M. That means that the line bundle $[D]$ is trivial, so there exists an open cover $\{U_\alpha\}$ of M and holomorphic non-vanishing functions $f_\alpha : U_\alpha \to \mathbb{C}^*$ such that

$$\frac{f_\alpha}{f_b} = g_{\alpha\beta} = \frac{h_\alpha}{h_b} \qquad \text{on } U_\alpha \cap U_\beta,$$

where h_α is a local defining meromorphic function for D on U_α. This shows the existence of a global meromorphic function H on M such that $H\big|_{U_\alpha} = h_\alpha/f_\alpha$. Moreover, as f_α does not vanish on U_α, the divisor associated to H is just D. This proves the injectivity of $[\ ]$ on linear equivalence classes of divisors.

Now, every holomorphic line bundle of a projective manifold has a global meromorphic section (see [3], p. 161). If $L \in \mathrm{Pic}(M)$ is a holomorphic line bundle, we have seen that a global meromorphic section σ of L defines a divisor (σ) on M. We claim that $[(\sigma)] = L$. If $g_{\alpha\beta}$ denote the transition functions of L with respect to some trivialization (U_α, ψ_α), the meromorphic section σ defines meromorphic functions σ_α on U_α such that $g_{\alpha\beta} = \sigma_\alpha/\sigma_\beta$. From the definition, σ_α is a defining meromorphic section for (σ) on U_α, thus L is just the line bundle associated to (σ). Summarizing, we have proved the following:

THEOREM 24.3. *If the manifold M is projective, the homomorphism $[\ ]$ descends to an isomorphism*

$$(\mathrm{Div}(M)/_{\equiv}) \xrightarrow{\cong} \mathrm{Pic}(M).$$

24.3. Adjunction formulas

Let $V \subset M$ be a smooth complex hypersurface of a compact complex manifold M. We will show that the normal and conormal bundles of V in M can be computed in terms of the line bundle associated to the divisor V.

PROPOSITION 24.4. (First adjunction formula) *The holomorphic normal bundle of V in M is isomorphic to the restriction to V of the line bundle $[V]$:*

$$N_V = [V]\big|_V.$$

PROOF. Let $i : V \to M$ be the inclusion of V into M. By definition, the normal bundle N_V is the cokernel of the inclusion $i_* : T^{1,0}V \to T^{1,0}M\big|_V$ and its dual, the conormal bundle N_V^*, is defined as the kernel of the projection $i^* : \Lambda^{1,0}M\big|_V \to \Lambda^{1,0}V$. Thus N_V^* is spanned by holomorphic $(1,0)$-forms on M vanishing on V.

Let f_α be local defining functions for V on some open covering U_α. By definition, the quotients $g_{\alpha\beta} := f_\alpha/f_\beta$ are the transition functions of $[V]$ on $U_\alpha \cap U_\beta$. Moreover, since f_α vanishes along V which is smooth, we see that $df_\alpha\big|_V$ is a non-vanishing local section of N_V^*. Now, since $f_\alpha = g_{\alpha\beta}f_\beta$, we get

$$df_\alpha\big|_V = (f_\beta dg_{\alpha\beta} + g_{\alpha\beta}df_\beta)\big|_V = g_{\alpha\beta}\big|_V df_\beta\big|_V.$$

Thus the collection $(U_\alpha, df_\alpha\big|_V)$ defines a global holomorphic section of $N_V^* \otimes [V]\big|_V$, showing that this tensor product bundle is trivial. This proves that $N_V = [V]\big|_V$. $\qquad\square$

We now denote by K_M and K_V the canonical line bundles of M and V. Consider the exact sequence of holomorphic vector bundles over V:

$$0 \to N_V^* \to \Lambda^{1,0}M\big|_V \to \Lambda^{1,0}V \to 0.$$

Taking the maximal exterior power in this exact sequence yields

$$K_M\big|_V \simeq K_V \otimes N_V^* = K_V \otimes [-V]\big|_V,$$

so

$$K_V \simeq (K_M \otimes [V])\big|_V.$$

This is the *second adjunction formula*.

We will use the following theorem whose proof, based on the Kodaira vanishing theorem, can be found in [3], p. 156.

THEOREM 24.5. (Lefschetz Hyperplane Theorem) *Let V be a smooth analytic hypersurface in a compact complex manifold M^{2m} such that $[V]$ is positive. Then the linear maps $H^i(M, \mathbb{C}) \to H^i(V, \mathbb{C})$ induced by the inclusion $V \to M$ are isomorphisms for $i \leq m - 2$ and injective for $i = m - 1$. If $m \geq 3$ then $\pi_1(M) = \pi_1(V)$.*

Our main application will be the following result on complete intersections in the complex projective space.

THEOREM 24.6. *Let k be a positive integer and let $P_1, \ldots, P_k$ be homogeneous irreducible relatively prime polynomials in $m + 1$ variables of degrees $d_1, \ldots, d_k$. Let N denote the subset in $\mathbb{CP}^m$ defined by these polynomials:*

$$N := \{[z_0 : \ldots : z_m] \in \mathbb{CP}^m \mid P_i(z_0, \ldots, z_m) = 0, \ \forall \ 1 \leq i \leq k\}.$$

Then, if N is smooth, we have $K_N \simeq [qH]\big|_N$, where $q = (d_1 + \cdots + d_k) - (m+1)$ and H is the hyperplane divisor in $\mathbb{CP}^m$.

PROOF. Notice first that N is smooth for a generic choice of the polynomials P_i. We denote by V_i the analytic hypersurface in $\mathbb{CP}^m$ defined by P_i and claim that

$$V_i \cong d_i H. \tag{24.1}$$

This can be seen as follows. While the homogeneous polynomial P_i is not a well-defined function on $\mathbb{CP}^m$, the quotient $h_i := P_i/z_0^{d_i}$ is a meromorphic function. More precisely, h_i is defined by the collection $(U_\alpha, P_i/z_\alpha^{d_i}, z_0^{d_i}/z_\alpha^{d_i})$. Clearly the zero-locus of h_i is $(h_i)_0 = V_i$ and the pole-locus is $(h_i)_\infty = d_i H_0$, where H_0 is just the hyperplane $\{z_0 = 0\}$. This shows that $(h_i) = V_i - d_i H_0$, thus proving our claim.

For $i = 1, \ldots, k$, let N_i denote the intersection of $V_1, \ldots, V_i$. Since $N_{i+1} = N_i \cap V_{i+1}$, we have

$$[N_{i+1}]\big|_{N_i} = [d_{i+1}H]\big|_{N_i}. \tag{24.2}$$

This follows from the fact that if V is an irreducible hypersurface in a projective manifold M and N is any analytic submanifold in M then

$$[V]\big|_N \simeq [V \cap N].$$

We claim that

$$K_{N_i} \simeq [n_i H]\big|_{N_i}, \qquad (24.3)$$

where $n_i := (d_1 + \cdots + d_i) - (m+1)$. For $i = 1$ this follows directly from the second adjunction formula together with (24.1), using the fact that $K_{\mathbb{CP}^m} = [-(m+1)H]$ (by (9.5)). Suppose that the formula holds for some $i \geq 1$. The second adjunction formula applied to the hypersurface N_{i+1} of N_i, together with (24.2) yields

$$K_{N_{i+1}} = ([N_{i+1}] \otimes K_{N_i})_{N_{i+1}} = ([d_{i+1}H] \otimes [n_i H])\big|_{N_{i+1}} = [n_{i+1}H]\big|_{N_{i+1}}.$$

Thus (24.3) is true for every i, and in particular for $i = k$. This finishes the proof. ☐

COROLLARY 24.7. *Let $d_1, \ldots, d_k$ be positive integers and denote their sum by $m+1 := d_1 + \cdots + d_k$. Suppose that $m \geq k+3$. If $P_1, \ldots, P_k$ are generic homogeneous irreducible polynomials in $m+1$ variables of degrees $d_1, \ldots, d_k$, then the manifold*

$$N := \{[z_0 : \ldots : z_m] \in \mathbb{CP}^m \mid P_i(z_0, \ldots, z_m) = 0, \ \forall \ 1 \leq i \leq k\}$$

carries a unique (up to rescaling) Ricci-flat Kähler metric compatible with the complex structure induced from $\mathbb{CP}^m$. Endowed with this metric, N is Calabi–Yau.

PROOF. Theorem 24.6 shows that the first Chern class of N vanishes. Since $m \geq k+3$, the Lefschetz Hyperplane Theorem applied inductively to the analytic hypersurfaces $N_i \subset N_{i+1}$ shows that N is simply connected and $b_2(N) = b_2(\mathbb{CP}^m) = 1$. Moreover the restriction of the Kähler form of $\mathbb{CP}^m$ to N is a generator of the second cohomology group of N. By the Calabi–Yau theorem, there exists a unique Ricci-flat metric on N up to rescaling. If this metric were reducible, we would have at least two independent elements in the second cohomology of N, defined by the Kähler forms of the two factors. Since $b_1(N) = 1$ this is impossible. Thus N is either Calabi–Yau or hyperkähler. The latter case is however impossible, since every compact hyperkähler manifold has a non-trivial parallel $(2, 0)$-form, thus its second Betti number is at least 3. ☐

24.4. Exercises

(1) Let $F_m(k)$ be the Fermat hypersurface of degree k in $\mathbb{CP}^m$:

$$F_m(k) := \{[z_0 : \ldots : z_m] \in \mathbb{CP}^m \mid \sum_{j=0}^{m} z_j^k = 0\}.$$

Prove that $F_m(k)$ is a smooth complex manifold. Show that $F_m(k)$ carries a Kähler–Einstein metric with negative scalar curvature if and only if $k \geq m + 2$.

(2) Show that every smooth cubic surface in $\mathbb{CP}^2$ is diffeomorphic to a 2-torus. *Hint:* Use Theorems 24.6 and 18.1. Conclude by Proposition 6.7.

(3) Let $L = M \times \mathbb{C}$ be the trivial complex line bundle on a compact complex manifold M. Show that L has a non-trivial holomorphic structure if and only if $H^{0,1}M \neq 0$. *Hint:* If $\bar{\partial}$ is a non-trivial holomorphic structure on L, and s is a nowhere zero section of L, one can write $\bar{\partial}s = \alpha \otimes s$ for some $\alpha \in \Omega^{0,1}M$. Since $\bar{\partial}^2 = 0$ one has $\bar{\partial}\alpha = 0$. The cohomology class of α in $H^{0,1}M$ is non-zero since otherwise one could construct a nowhere zero holomorphic section of L. The converse is similar.

(4) Show that every complex line bundle on $\mathbb{CP}^m$ carries a unique holomorphic structure up to isomorphism. Deduce that the Picard group of $\mathbb{CP}^m$ is isomorphic to $\mathbb{Z}$.

Bibliography

[1] A. Besse, *Einstein Manifolds*, Ergebnisse der Mathematik und ihrer Grenzgebiete, **10** (Berlin: Springer, 1981).

[2] P. Gauduchon, *Calabi's Extremal Kähler Metrics* (in preparation).

[3] P. Griffith and J. Harris, *Principles of Algebraic Geometry* (New York: Wiley, 1978).

[4] M. Gromov and B. Lawson, Jr., *The classification of simply connected manifolds of positive scalar curvature. Ann. Math.* **111** (1980), 423–434.

[5] F. Hirzebruch, *Topological Methods in Algebraic Geometry* (New York: Springer, 1966).

[6] L. Hörmander, *An Introduction to Complex Analysis in Several Variables* 3rd edn (Amsterdam: North-Holland, 1990).

[7] D. Joyce, *Compact Manifolds with Special Holonomy* (Oxford: Oxford University Press, 2000).

[8] E. Kähler, *Über eine bemerkenswerte Hermitesche Metrik. Abh. Math. Sem. Hamburg Univ.* **9** (1933), 173–186.

[9] S. Kobayashi, *On compact Kähler manifolds with positive definite Ricci tensor. Ann. Math.* **74** (1961), 570–574.

[10] S. Kobayashi and K. Nomizu, *Foundations of Differential Geometry I* (New York: Interscience Publishers, 1963).

[11] S. Kobayashi and K. Nomizu, *Foundations of Differential Geometry II* (New York: Interscience Publishers, 1969).

[12] S. Kobayashi, *Differential Geometry of Complex Vector Bundles* (Princeton, NJ: Princeton University Press, 1987).

[13] S. Lang, *Differential Manifolds* (Reading, MA: Addison-Wesley, 1972).

[14] A. Lichnerowicz, *Spineurs harmoniques. C. R. Acad. Sci. Paris* **257** (1963), 7–9.

Index

Symbols

$i\partial\bar{\partial}$-lemma
  global -, 109
  local -, 68

A

adjunction formulas, 162, 163
almost
  complex manifold, 60
  complex structure, 57, 60
  Hermitian metric, 57
analytic hypersurface, 159
associated vector bundle, 34
atlas, 3
Aubin–Yau theorem, 129

B

Berger holonomy theorem, 154
Betti numbers, 106
Bianchi identities, 50
Bochner formula, 139
bump function, 7

C

Calabi–Yau
  manifold, 144
  theorem, 125
canonical bundle, 74
Cartan formula, 24
Cauchy–Lipschitz theorem, 9
Čech
  cocycle, 31
  cohomology, 32
chain rule, 4
chart, 3
Chern
  class, 113
  connection, 79
Chern–Weil theory, 113
Christoffel symbols, 89

closed exterior form, 23
coclosed exterior form, 101
codifferential, 101
complete vector field, 9
complex
  manifold, 59
  projective space, 59
  structure, 62
  vector bundle, 71
  volume form, 121
connection, 40
  $\mathbb{C}$-linear -, 78
  form, 45
contraction, 13
cotangent bundle, 15
covariant derivative, 37
curvature
  operator, 77
  tensor, 50

D

diffeomorphism, 4
differential
  exterior -, 22
  of a map, 4, 5
distribution, 26
divisor, 159
Dolbeault
  cohomology, 107
  decomposition of a form, 108
  decomposition theorem, 108
  isomorphism, 108
  lemma, 68
duality
  Kodaira–Serre -, 145
  Poincaré -, 106
  Serre -, 108

E

effective divisor, 159
Einstein metric, 50
enlargement of the structure group, 32
Euler characteristic, 143
exact exterior form, 23
exterior
  bundle, 21
  derivative, 22
  form, 13
  product, 13

F

fibre, 31
flat Riemannian metric, 50
formal adjoint, 99
frame bundle, 32
free action, 31
Fubini–Study metric, 93
fundamental 2-form, 81

G

G-structure, 31
  geometrical -, 119
  topological -, 119

H

harmonic form, 101, 105
Hermitian
  connection, 79
  metric, 81
  structure, 78
  vector bundle, 78
Hirzebruch–Riemann–Roch formula, 143
Hodge
  ∗-operator, 100
  decomposition of a form, 106
  decomposition theorem, 105
  isomorphism, 106
  numbers, 108
holomorphic
  Euler characteristic, 143
  form, 67
  function, 59
  map, 59
  section, 72
  structure
    on manifolds, 59
    on vector bundles, 72
  symplectic form, 156
  vector bundle, 71
  vector field, 67

holonomy
  bundle, 44
  group, 43
horizontal
  distribution, 39
  lift, 39, 42
  path, 42
  section, 40
hyperkähler manifold, 153
hyperplane
  divisor, 161
  line bundle, 76, 155, 161

I

infinitesimal isometry, 51
integrable
  almost complex structure, 57, 69
  distribution, 26
integral curve, 8
interior product, 14
irreducible
  analytic hypersurface, 159
  Riemannian manifold, 49
isometry, 51

K

Kähler
  –Einstein metric, 129
  form, 57
  identities, 102
  metric, 82
  potential, 58
  structure, 57
Killing vector field, 51
Kodaira
  embedding theorem, 155
  vanishing theorem, 151
Kodaira–Serre duality, 145

L

Laplace operator, 101
Lefschetz Hyperplane Theorem, 163
Levi–Civita connection, 49
Lichnerowicz formula, 147
Lie
  algebra, 29
  bracket, 17
  derivative, 17
  group, 29
linear connection, 41
linearly equivalent divisors, 161
local

chart, 3
connection forms, 77
coordinate system, 4
curvature forms, 77
defining function, 159, 160
flow, 9
frame, 6
Kähler potential, 82

M

Matsushima theorem, 133
meromorphic
   function, 159
   section, 160

N

Newlander–Nirenberg theorem, 61
Nijenhuis tensor, 66
normal coordinates, 58

O

orientation, 4, 32

P

parallel
   section, 38
   transport, 43
partition of unity, 25
Picard group, 161
Poincaré duality, 106
positive
   $(1, 1)$-form, 125, 141
   first Chern class, 129
   line bundle, 141
principal bundle, 31
projective manifold, 155
projective space, 9, 59
pseudo-holomorphic structure, 72
pull-back
   bundle, 41
   connection, 42
   covariant derivative, 42
   of a covariant tensor, 16
push forward of a tensor, 16

R

real holomorphic vector field, 67, 69
reducible connection, 44
reduction of the structure group, 32
restricted holonomy group, 43
Rham, de
   cohomology, 105

decomposition theorem, 154
isomorphism theorem, 105
Ricci
   form, 87, 116
   tensor, 50
Ricci-flat metric, 51
Riemannian
   curvature tensor, 50
   manifold, 47
   metric, 47
   submersion, 97

S

section
   of a principal bundle, 31
   of a vector bundle, 33
Serre duality, 108
sheaf, 155
smooth
   manifold, 4
   map, 4
   structure, 4
Stokes theorem, 25
submanifold, 5
submersion theorem, 5

T

tangent
   bundle, 5
   space, 5
   vector, 5
tautological line bundle, 74
tensor
   algebra, 13
   bundle, 15
   field, 15
topological
   G-structure, 119
   manifold, 3
torsion, 41
transitive action, 31

V

vector
   bundle, 33
   field, 6
volume form, 49
   complex -, 121

W

Weitzenböck formula, 135
Whitney sum, 113

Printed in the United States
By Bookmasters